Smart Sensing with Digital Twins

This book explores the innovative use of small unmanned aircraft systems (sUAS)—commonly known as drones—for methane emissions monitoring (i.e., detection, localization, and quantification) by introducing smart sensing frameworks and digital twin technology.

Based on the concept of smart sensing, which combines mobile sensor data with physics-based models to provide actionable and timely insights, this book presents novel methods for monitoring and quantifying methane emissions, a potent greenhouse gas, using digital twins in single and multiple sUAS-based approaches. The first part of the book examines the methane sensing problem for detecting, locating, and quantifying emission sources, with case studies highlighting key observations and lessons learned from field experiments. The second part proposes what, why, and how digital twins should be used in environmental monitoring applications, covering both basic detection principles and advanced source localization and quantification techniques. This section shows how digital twins can enhance sUAS-based source detection and smart sensing methods.

With practical tools and field-tested examples, this book serves as both an introductory guide and advanced reference to environmental monitoring and is particularly valuable to researchers, students, engineers, and environmental professionals in engineering, environmental studies, and technology.

Derek Hollenbeck earned his Ph.D. in Mechanical Engineering in 2023 at the University of California, Merced. His research interests include digital twins, fluid mechanics, controls, fractional calculus, and cyber-physical systems using small unmanned aerial systems with applications in methane leak detection and quantification.

YangQuan Chen is a professor at the University of California, Merced, US. His research interests include mechatronics for sustainability, digital twins, small multi-UAV, and applied fractional calculus. His recent publications with CRC Press include *Fractional Calculus for Skeptics I: The Fractal Paradigm*.

Smart Sensing with Digital Twins

Methane Emission Source Determination
with sUAS

Derek Hollenbeck and YangQuan Chen

CRC Press
Taylor & Francis Group
Boca Raton London New York

CRC Press is an imprint of the
Taylor & Francis Group, an **informa** business

Designed cover image: Derek Hollenbeck

This book was supported under the University of California Merced's Climate Action Seed Funds grant for the Center for Methane Emission Research and Innovation.

MATLAB® and Simulink® are trademarks of The MathWorks, Inc. and are used with permission. The MathWorks does not warrant the accuracy of the text or exercises in this book. This book's use or discussion of MATLAB® or Simulink® software or related products does not constitute endorsement or sponsorship by The MathWorks of a particular pedagogical approach or particular use of the MATLAB® and Simulink® software.

First edition published 2026
by CRC Press
2385 NW Executive Center Drive, Suite 320, Boca Raton FL 33431

and by CRC Press
4 Park Square, Milton Park, Abingdon, Oxon, OX14 4RN

CRC Press is an imprint of Taylor & Francis Group, LLC

ISBN: 978-1-041-13229-5 (hbk)
ISBN: 978-1-041-13385-8 (pbk)
ISBN: 978-1-003-66947-0 (ebk)

DOI: 10.1201/9781003669470

Typeset in Latin Modern font
by KnowledgeWorks Global Ltd.

Publisher's note: This book has been prepared from camera-ready copy provided by the authors.

To Elvira, my family, & mentors,
without your support this accomplishment would not have been
possible.

To Dalia,
through smart thinking, hard work, and persistent effort,
anything is possible.

To my readers,
I hope this book helps you implement smart sensing systems to
make the world a better place.

— D.H.

To my mentors and family.

— Y.C.

Contents

Preface

During our time as graduate students, when it came to topics of research, we had always been tasked with solving the problem of "so what, who cares, and why you". Therefore, our research into the use of small unmanned aircraft systems (sUAS – also referred to as drones) with the methane emission source determination problem has been largely motivated by climate change and the warming of our planet. Methane is a very potent greenhouse gas compared to carbon dioxide and it has high potential to bring change to global warming effects due to its decreased atmospheric lifespan – i.e. our 'control' knob for near-term change. However, it turns out, methane is highly challenging to measure. It is a colorless and odorless gas requiring special sensors to detect. Therefore, we focused our efforts on solving this problem because it has the potential of real-world impact and is difficult to solve. However, before we can achieve this goal, we need to fully understand the measurement problem. This will ultimately provide pathways for detecting leaks, locating their sources, and quantifying the emissions, but more importantly it provides pathways for measurement, verification (i.e. confirming that they have been fixed), and reporting programs. As Peter Drucker said, "what gets measured, gets managed".

In the literature, in the context of applications, there has been discrepancies between the top-down (or direct measurements) and the bottom-up (or inventory-based) approaches. Often the bottom-up approach under estimates the emission source, which prompts investigation into improving the emission factors associated with these kinds of methods. There are also issues with minimum detection limit and how industry is looking toward probability of detection to help understand how 'well' we can measure in all environments and scenarios – not just in 'ideal' cases. Understanding how different sensing technologies and how the different modes of measuring methane (in situ, path-integrate, imaging-based) are effective in localization and quantification tasks are key to addressing this issue.

In the past decade, we have seen the integration of Digital Twins (DT) into key research areas (such as manufacturing and smart cities), improving the way we provide solutions. Currently, DTs have not been applied to environmental systems research, especially in the source determination problem. The benefits from DT technology success in other industries give insights on how they may be useful in this problem. Furthermore, advances in industry 4.0 thinking or advances in IoT and edge devices with DTs provide the potential for making methane sensing smarter within all sorts of applications – oil and gas, agriculture, dairy, landfill, and even permafrost. Within the DT framework, a vision can be formed for how these measurements can be automated for more robust sensing, early detection, and faster repairs.

Many of the techniques employed by practitioners rely on satellite, manned aircraft, foot-based measurements, or by installing many fixed sensors (also referred to as continuous emission monitoring). However, it is wrong to say that one approach is a 'one size fits all' and better than the others. A much better approach is to provide a holistic view of the measurement problem by providing complementary sensing at different measurement scales. And, there happens to be a gap between ground measurements and manned aircraft that is suitable for drone-based sensing – from 400 feet above ground level (AGL) to the surface. Drones also offer a unique capability for being deployed; they have the ability to be re-configured with different sensors, they can access areas between the ground and manned aircraft that traditional vehicles or personnel on foot cannot access (e.g. over bodies of water or tree-lines, or next to flares or tanks), they can provide faster and more frequent surveys with finer spatial resolutions, and they can be relatively more cost effective than traditional surveys.

The atmospheric turbulence and weather pose challenges in the use of DTs and drone systems, both for flight capabilities and for solving the emission source determination in general. For example, strong winds and extreme cold/hot conditions can prevent or reduce flight times. Or, specific weather conditions may prohibit a sensor from its use (e.g. in rain or high humidity conditions). Additionally, as the terrain becomes more complex, the air flow around that terrain also becomes more complex, requiring more sophisticated DT modeling to capture higher levels of detail (or fidelity). This higher fidelity results in an increase in the dimensionality of the modeling problem. Since environmental modeling can already be costly to undertake – high dimensionality – added complexities from terrain and weather further increases the computational expense for the DT, limiting and or preventing the useful real-time capabilities. Of which is a core goal of smart sensing and this book.

This book offers a new way to look at the environmental sensing problem with sUAS by incorporating DTs and smart sensing frameworks. First, we integrate hybrid style modeling, allowing for near real-time computation, with advanced leak detection and quantification methods. Second, we investigate 'where to sense' through integrating multiple sUAS and quantified observability concepts. The combination of the two enables smarter measurements and even allows for more simulation-based testing to avoid unnecessary field work and controlled release testing (speeding up the development iteration cycle).

This book is designed and written for graduate students, researchers, scientists, field operators/drone specialists, industry experts, and policy makers in an introductory fashion. The book contains both high-level explanations and examples, as well as mathematical details for a general understanding. For the interested reader, the references included therein, are sufficient to fully understand and implement the techniques. To help foster the use of the topics covered in this book, a Github repository will be included, such that the reader can explore and play with the proposed DT framework. Unfortunately, due to non-disclosure agreements, not all of the data used to produce the results of this manuscript will be made publicly available. However, some of the controlled release data will be made available for use and reference.

The book is divided into two sections. The first section will cover an introduction to the methane sensing problem, and the detection, localization, and quantification of methane emission sources with sUAS. The aim is to provide a general overview of these topics and give the important aspects related to each subproblem. Within each chapter, a 'Pause and Reflect' is included to help the reader think about how these topics can be expanded and also prompts the readers thinking for the next chapter section or chapter. A series of case studies is showcased at the end of the first section, highlighting the experiments done, but more importantly, the observations made and lessons learned during the field experiments. The second section will introduce the concepts of DTs and their use cases (include some case studies), as well as build-up the concept of smart sensing, sensor placement, and steering problem. The goal of this section is to highlight how DTs can be leveraged to expand sUAS-based source determination problem method developments, but also embed smartness into the 'how to best sense' questions, with respect to observability or solving the inverse problem. At the end, we highlight some case studies on the topic of smart sensing before concluding with a summary of the book's takeaways, lessons learned, and the MOABS/DT code breakdown.

We wish to thank the funding support by the Center for Methane Emission Research and Innovation (CMERI) through the Climate Action Seed Funds grant (2023–2026) at the University of California, Merced. We also wish to thank MESA Lab members and many undergraduate researchers involved in various field campaigns. We also wish to thank Dr. Lance Christensen of JPL for jointly starting the methane drone detection project in 2014. This book grows out of the first author's 2023 Ph.D. dissertation with some new yet systematic developments included.

Last but not least, the authors would like to thank Ms Xiaoyin Feng, Books Editorial Assistant, Routledge & CRC Press, Taylor & Francis Group, for her professional coordination of the book project; Ms. Lian Sun, Publisher, Head of China Books Publishing and International Cooperation Taylor & Francis Advanced Learning for her professional vision and help in the peer review process.

CA USA, April 2025   *Derek Hollenbeck*
CA USA, April 2025   *YangQuan Chen*

List of Acronyms

ADE	Advection diffusion equation
ADMM	Alternating direction method of multipliers
AGL	Above ground level
AMFC	Alberta Methane Field Challenge
APEX	Alaskan Peatland Experiment
APRA-E	Advanced Research Projects Agency-Energy
AVIRIS	Airborne Visible/Infrared Imaging Spectrometer
AVIRIS-NG	Next Generation AVIRIS
BFGS	Broyden-Fletcher-Goldfarb-Shanno
bLS	Backwards Lagrangian stochastic
Bm	Brownian motion
CARB	California Air Resources Board
CFD	Computational fluid dynamics
CFL	Courant-Friedrichs-Lewy
CFP	Cylindrical flux plane
CFR	Code of Federal Regulations
CMERI	Center for Methane Emission Research and Innovation
CMI	Concentration measurement instrument
CRDS	Cavity ring-down spectrometer
CRF	Controlled release facility
CRLB	Cramer Rao lower bound
CVT	Centroidal Voronoi tessellations
D-CVT	Density CVT

DE-CVT	Density and entropy CVT
DMD	Dynamic mode decomposition
DM+V	Distribution modeling with variance
DM+V/W	Distribution modeling with variance and wind
DT	Digital twin
EC	Eddy covariance
ECMWF	European Centre for Medium-Range Weather Forecasts
EDF	Environmental Defense Fund
EMI	Electromagnetic interference
EPA	Environmental Protection Agency
ESC	Extremum seeking control
fBm	Fractional Brownian motion
FFT	Fast Fourier transform
fGn	Fractional Gaussian noise
FID	Flame ionization detector
FIM	Fisher information matrix
FISTA	Fast iterative shrinking-threshold algorithm
FNO	Fourier neural operators
FOM	Figures of merit
FTIR	Fourier transformed infrared
GCS	Ground control station
GCV	Generalized cross-validation
GDT	Gauss divergence theorem
GHG	Greenhouse gas
GLM-VFP	general linear model VFP
Gn	Gaussian noise
GPM	Gaussian plume model
GPS	Global positioning system

GWP	Global warming potential
HOT	High operating temperature
IDW	Inverse distance weighting
IMAP-DOAS	Iterative maximum a posterior differential optical absorption spectroscopy method
IME	Integrated mass enhancement
IoT	Internet of things
IR	Infrared
ISTA	Iterative shrinking-threshold algorithm
JPL	Jet Propulsion Lab
LASSO	Least absolute shrinkage and selection operator
LDAQ	Leak detection and quantification
LDAR	Leak detection and repair
LGD	Laser gas detection
LGR	Los Gatos Research
LiDAR	Light detection and ranging
LTA	Lighter than air
LWIR	Long-wave infrared
MB	Mass balance
MCMC	Markov chain Monte Carlo
MCRWM	Merced County Regional Waste Management
METEC	Methane emission technology evaluation center
MGGA	micro-portable greenhouse gas analyzer
MLE	Maximum likelihood estimator
MMC	Mobile Monitoring Challenge
MMD	Micrometeorological mass difference
MMF	Modeled mass flux
MOABS/DT	Methane Odor Abatement Simulator Digital Twin

MONITOR	Methane Observation Networks with Innovative Technology to Obtain Reductions
MOS	Metal oxide sensors
MOST	Monin-Obukhov similarity theory
MSWL	Municipal solid waste landfill
MUST	Mock Urban Setting Test
MVPGR	Merced Vernal Pools and Grassland Reserve
MWIR	Mid-wave infrared
NASA	National Aeronautics and Space Administration
NGI	Near-field Gaussian plume inversion
NIR	Near infrared
NSF	National Science Foundation
ODE	Ordinary differential equation
OGI	Optical gas imaging
OGMP	Oil and Gas Methane Partnership
OPLS	Open path laser spectrometer
OTM	Other test method
PD	proportional derivative
PDE	Partial differential equation
PE	Persistence of excitation
PFP	Perimeter flight plan
PG	Pasquill-Gifford
PGE	Pacific Gas & Electric
PIML	Physics-informed machine learning
PINN	Physics-informed neural network
PI-RNN	Physics-informed recurrent neural network
PI-VFP	Path integrated VFP
pMGGA	prototype micro-portable greenhouse gas analyzer

POD	Probability of detection
PSG	Point source Gaussian
PSG-CS	Conditionally sampled PSG
PSG-RB	Recursive Bayesian PSG
PSG-SBM	Sequential Bayesian MCMC PSG
PSI	pounds per square inch
QCLS	Quantum cascade laser spectrometer
QOGI	Quantitative OGI
QUIC	Quick Urban and Industrial Complex
RBF	Radial basis function
RF	Radio frequency
RMLD	Remote methane leak detector
RNN	Recurrent neural networks
SCFH	Standard cubic feet per hour
SciML	Scientific machine learning
SDE	Stochastic differential equation
SDK	Software development kit
SEM	Surface emission monitoring
SOTA	State-of-the-art
SPSA	Simultaneous perturbation and stochastic approximation
SR	Sufficient richness
STILT	Stochastic time inverted Lagrangian transport
sUAS	small unmanned aircraft system
SVD	Singular value decomposition
TCM	Tracer correlation method
TDLAS	Tunable diode laser absorption spectrometer
TI	Turbulence intensity
TVA	Toxic vapor analyzers

UGGA   Ultra-portable greenhouse gas analyzer

UNEP   United Nations Environment Programme

USR   Upwind survey region

VFP   Vertical flux plane

VOC   Volatile organic compounds

VRPM   Vertical radial plume mapping

VTOL   Vertical takeoff and landing

WRF   Weather research and forecasting

I

From Detection to Quantification: sUAS-Based Methane Sensing Techniques

CHAPTER 1

The Methane Sensing Problem

A Problem well defined is a problem half solved. In the first part of this book, we will go through the motivations behind why we are focusing on methane, how we detect, localize, and quantify methane using small unmanned aircraft systems (sUAS), and explore some case studies that help shape our understanding of this problem.

1.1 WHY METHANE?

Methane is a molecule comprising of a single carbon atom and four hydrogen atoms, denoted as CH_4. The global warming potential (with respect to CO_2) is referenced as 25 in a 100-year timespan and over 85 in a 20-year timespan. This means that CH_4 is a potent greenhouse gas (GHG) and contributes to the overall climate change problem.

What is climate change? The United Nations defines it as "...long-term shifts in temperatures and weather patterns." These shifts can be caused by natural forces, such as the sun's activity or through large atmospheric disturbances (e.g. volcanic eruptions) or from human activities, such as the burning of fossil fuels including power generation, manufacturing goods, or deforestation. According to several sources the human factor contributes to over 100% of the observed climate change over the past 50 years. This is in part due to some cooling associated with decreased insolation from the sun. However, there are still some questions about how natural ecosystems play a role in the overall dynamics of climate change on longer time scales. For example, the thawing of the arctic permafrost leads to the release of carbon dioxide and methane – contributing to more warming of the atmosphere.

How do we reduce or mitigate the effects of GHG warming? What is our control knob? The answer is methane. This is due to the atmospheric lifespan of CO_2 being on the order of 100 years. This means that CO_2 emitted today will survive and contribute to the greenhouse effect for the next 100 years. On the other hand, CH_4 has an atmospheric lifespan of around 10 years. The advantage is that by mitigating unnecessary CH_4 emissions today, we can see global impacts in the near term.

DOI: 10.1201/9781003669470-1

FIGURE 1.1 (top left) An laser spectrometer-based apparatus at UC Davis for measuring enteric methane emissions from cows [8]. (bottom left) A landfill gas collection system [1]. (right) A series of pump jacks and a lit flare burning methane [2].

1.2 KEY EMISSION SOURCES

The U.S. Environmental Protection Agency (EPA) outlines key anthropogenic (human-caused) emission sources in its annual Inventory of U.S. Greenhouse Gas Emissions and Sinks report. Major emission sources include fossil fuel combustion for electricity generation, transportation, and industrial activities, which together account for the majority of carbon dioxide (CO_2) emissions. Methane (CH_4) emissions primarily come from agriculture (notably enteric fermentation in livestock), landfills, and the oil and gas sector (see Fig. 1.1). The industrial sector also contributes significant nitrous oxide (N_2O) and fluorinated gases from chemical production and other manufacturing processes. This report emphasizes the ongoing need for mitigation strategies to address these emissions, particularly from energy and agricultural sectors.

Methane (CH_4) is a critical emission source to focus on because it is a highly potent greenhouse gas with a global warming potential (GWP) far greater than carbon dioxide (CO_2) over short time periods. Research by Drew Shindell and others has highlighted that methane's GWP is about 84–87 times stronger than CO_2 over a 20-year period, making it a key driver of near-term climate change [10]. Shindell's work also emphasizes that targeting methane can yield immediate benefits, not only in slowing global warming but also in improving air quality, as methane contributes to the formation of ground-level ozone, a harmful pollutant. Reducing methane emissions can therefore offer dual climate and public health benefits while providing a quicker return on mitigation efforts compared to CO_2.

Natural methane emission sources include wetlands, oceans, and permafrost, with wetlands being the largest contributor due to anaerobic (oxygen-deprived) decomposition of organic material. Climate change can exacerbate these natural emissions through positive feedback loops. As global temperatures rise, permafrost in the Arctic and sub-Arctic regions begins to thaw, releasing vast amounts of methane previously trapped in frozen organic material. Additionally, warmer conditions increase microbial activity in wetlands, further boosting methane emissions. This release of methane—one of the most potent greenhouse gases—amplifies the greenhouse effect, which in turn accelerates warming, creating a dangerous feedback cycle that can make controlling global temperatures even more difficult. This methane-climate feedback poses a significant risk for worsening global warming beyond current predictions.

Recent policies aimed at reducing methane emissions have gained momentum, particularly in the energy and agricultural sectors. In 2021, the Global Methane Pledge was launched at COP26, where over 100 countries committed to reducing methane emissions by 30% by 2030 compared to 2020 levels. In the United States, the EPA's 2023 methane rule targets methane leaks from oil and gas operations, requiring stricter monitoring, detection, and repair of leaks, along with new standards for venting and flaring. These policies focus on rapidly addressing methane to slow near-term climate warming and improve air quality.

The Oil and Gas Methane Partnership (OGMP) 2.0 is a voluntary initiative launched by the United Nations Environment Programme (UNEP) and industry stakeholders to drive significant methane emission reductions in the oil and gas sector. It sets a new reporting framework for companies, requiring detailed and accurate methane emission data across the entire value chain, with the aim of achieving a 45% reduction in methane emissions by 2025 and a 60–75% reduction by 2030.

In the United States, the EPA's 0000a and 0000b rules are critical regulatory efforts under the Clean Air Act targeting methane emissions from new, modified, and existing oil and gas sources. The Quad Oa (0000a) rule, established in 2016, introduced performance standards for new and modified oil and gas sources, while the Quad Ob (0000b) rule, introduced in 2021, expands these standards to existing sources, focusing on leak detection, monitoring, and repair requirements. These rules are central to the U.S. strategy for reducing methane emissions from one of its largest sources.

Pause and Reflect

What kind of policies can be made to improve the mitigation of methane? How can these policies make it easier, financially, for owners and operators to make these mitigations and for measurement companies and technology makers to validate these reductions?

1.2.1   Why sUAS in Industry?

In the early days of oil and gas operations, leak detection relied on visual inspection and odor detection, where workers would identify leaks based on bubbling liquids or the characteristic smell of hydrocarbons. This method was highly subjective and ineffective for detecting invisible gases like methane. By the mid-20th century, soap bubble testing became common for pinpointing leaks in pipelines and valves, but it lacked the ability to quantify emissions. To address this limitation, the industry adopted Toxic Vapor Analyzers (TVA) and Flame Ionization Detectors (FID) in Leak Detection and Repair (LDAR) programs. These handheld devices allowed technicians to "sniff" around components to detect volatile organic compounds (VOC). However, this approach was labor-intensive, time-consuming, and only provided localized point measurements, making it impractical for monitoring large facilities.

For quantification, companies used the bagging technique, where a flexible enclosure was placed around leaking equipment, and the emission rate was calculated by measuring the gas accumulation over time. While effective for small, controlled leaks, bagging was impractical for large-scale monitoring and complex facilities due to the need for extensive manual labor and potential interference from environmental factors like wind. In the late 20th century, infrared gas correlation spectroscopy and fixed gas sensors were introduced, allowing for remote and proximal stationary monitoring. Aircraft-mounted infrared cameras provided a step forward, enabling large-area leak detection, but these methods suffered from low sensitivity, interference from environmental factors, and high operational costs. Modern advancements, such as Optical Gas Imaging (OGI), drone-mounted sensors, tunable diode laser absorption spectroscopy (TDLAS), and continuous monitoring networks, have revolutionized methane detection and quantification. These technologies offer real-time, high-resolution, and automated leak detection with greater accuracy. Unlike older methods, they provide wider spatial coverage, faster leak identification, and improved quantification capabilities, addressing the key challenges of traditional approaches.

More modern methane measurement approaches vary depending on the type of platform used, each presenting distinctive challenges that affect detection capability, spatial coverage, and quantification accuracy. Foot surveys and surface emission monitoring (SEM) using handheld sensors or backpack systems are limited by human mobility, making them time-consuming and difficult in inaccessible or hazardous areas. Additionally, environmental interference from wind fluctuations and terrain obstructions can impact measurement accuracy.

Surface Emission Monitoring (SEM)

Surface Emission Monitoring (SEM) is a technique used to detect and quantify methane emissions from ground-based sources, such as landfills, oil and gas sites, and industrial facilities. SEM is typically performed using mobile sensors mounted on ground vehicles, handheld instruments, or small uncrewed aerial systems (sUAS) [4, 3]). A key advantage of SEM is its ability to detect spatial variations in methane concentration across a surface, making it useful for identifying emission hotspots and leaks. SEM methods often involve traversing a predefined survey path while continuously measuring methane concentrations. These measurements can be processed using inverse modeling techniques to estimate emission fluxes.

Fixed monitoring using stationary sensor networks and fence-line systems provides continuous data collection but is constrained by limited spatial resolution. These systems require significant infrastructure investment, and prolonged exposure to harsh environmental conditions can lead to sensor drift and reduced accuracy over time. Some sensors also have slow response times, causing delays in real-time leak detection. Vehicle-based surveys, including mobile laboratories and car-mounted sensors, allow for efficient coverage of large areas. However, these methods face challenges such as road accessibility constraints, making it difficult to monitor off-road or industrial sites comprehensively. Additionally, wind variability during transit can distort concentration readings, and distinguishing emissions from multiple sources is complicated. Vehicle exhaust can also introduce background contamination, requiring careful calibration. Drone-based surveys (sUAS) provide high-resolution measurements, both spatial and temporal, with a range of sensor options including in situ, path-integrated, and imaging-based technologies. Despite their flexibility, sUAS's are constrained by limited battery life, payload restrictions, and sensitivity to atmospheric variability such as wind turbulence and thermal effects. Regulatory restrictions often limit flight operations, requiring special permissions for certain areas. Additionally, interpreting drone data for precise quantification demands sophisticated modeling techniques. Manned aircraft surveys using airborne spectrometers and hyperspectral imaging offer broad spatial coverage but come with high operational costs, requiring extensive resources for flight hours, personnel, and maintenance. These methods also have coarser spatial resolution, as aircraft must operate at high altitudes, making it challenging to detect small-scale sources. Weather conditions, such as turbulence and cloud cover, can further impact measurement accuracy, and post-processing of large datasets requires advanced algorithms to account for variations in altitude and wind conditions. Satellite-based observations provide a global perspective on methane emissions but have limitations in detection sensitivity and revisit frequency. Many satellites can only observe a given location every few days to weeks, making them less effective for real-time leak detection. Additionally, most satellite sensors have high detection limits, meaning only strong methane sources, known as super emitters, can be reliably identified. Cloud cover further reduces data reliability for optical and infrared sensors. The complexity of satellite data interpretation also requires sophisticated atmospheric models to separate methane signals from background noise.

Each method represents a trade-off between spatial coverage, temporal resolution, detection sensitivity, and operational feasibility. See Fig. 1.2 for an illustration of some of the methods discussed. In practice, a combination of these platforms are often used to achieve comprehensive methane monitoring program, integrating multiple data sources for more accurate emission detection and quantification. The interested reader should also check [7, 9, 6] for more details.

1.3 TYPES OF EMISSIONS

There are various types of emission sources found in applications and the key emission sources we just discussed. Emission types, such as a continuous point source, intermittent point source, uniform area source, distributed area source, elevated area source, or an underground point source (see Fig. 1.3). Although this list is not exhaustive, it covers the general types observed. Furthermore, the type of emission source can significantly impact the measurement, flight strategies, and sensor selection for

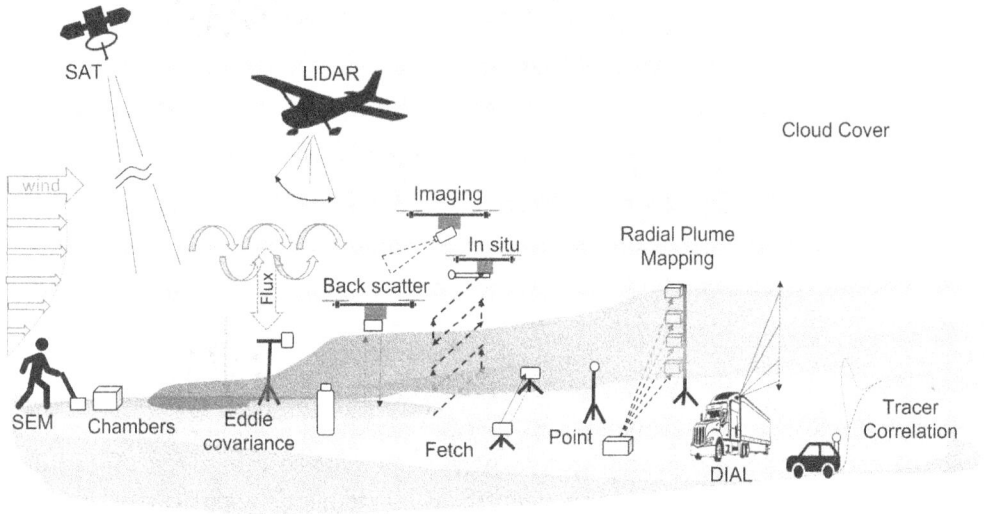

FIGURE 1.2 Examples of methane emission measurement and quantification approaches.

sUAS-based methane monitoring. The types of emission sources are described briefly here:

- **Continuous Point Source** – A fixed location emitting gas at a steady rate over time, such as a leaking pipeline valve or an industrial stack. These sources

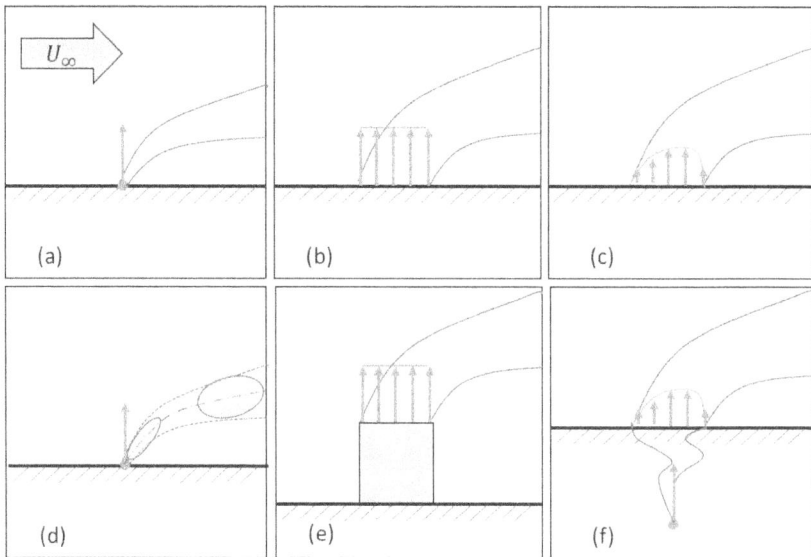

FIGURE 1.3 Example illustration for source types: (a) continuous point source, (b) uniform area source, (c) distributed area source, (d) intermittent point source, (e) elevated area source, and (f) underground point source. (reused from [6])

are well-suited for long-duration or repeated measurements with sUAS, as their emissions are relatively stable. Enabling the use of steady-state dispersion models and optimized flight paths to maximize data collection efficiency.

- **Intermittent Point Source** – A localized source that releases gas sporadically or cyclically, such as compressor station blowdowns or venting events. Capturing these emissions requires careful timing and rapid-response sUAS sensing to coincide with release events, often necessitating high-frequency sampling sensors and adaptable flight patterns.

- **Uniform Area Source** – A broad, evenly distributed emission zone, such as a landfill or wastewater treatment pond. These sources are challenging for sUAS-based quantification, as emissions may blend into background concentrations without distinct localized plumes. They pose challenges due to their broad spatial extent. Effective quantification requires grid-based flight patterns and high-sensitivity sensors to distinguish emissions from background concentrations.

- **Distributed Area Source** – A non-uniform emission zone with multiple localized sources contributing to the total flux, such as a well pad with multiple leaks. sUAS measurements must differentiate between individual sources and background noise, often requiring high-resolution spatial mapping.

- **Elevated Point/Area Source** – A source that releases emissions from an elevated surface, such as an oil storage tank, unlit flare, or biomass pile. These sources can create complex plume structures that require sUAS-based sampling at varying altitudes to fully characterize the emissions.

- **Underground Point Source** – A subsurface leakage point, such as a leaking pipeline buried beneath the soil. Detecting and quantifying these emissions with sUAS often relies on indirect indicators, such as localized concentration anomalies or surface flux measurements using spectral imaging or gas sensors capable of measuring close to the surface.

Each of these scenarios necessitates tailored flight strategies, sensor technologies, and data analysis methods to accurately detect, localize, and quantify methane emissions.

1.4 MEASUREMENT MODES AND TECHNOLOGIES

It is not enough to understand the importance of mitigating methane to be able to manage it. We also need to understand how we can sense or detect methane to make measurements. Like other atmospheric trace gases, methane has a unique absorption spectrum. When light interacts with methane gas, some of the energy is absorbed. This is the principal idea behind many measurement techniques. As we shall see in the following sections, applications of methane may require different modes of sensing to perform the task. Furthermore, we need to understand what technologies are available and practical to carry out these application dependent tasks.

METHANE
INFRARED SPECTRUM

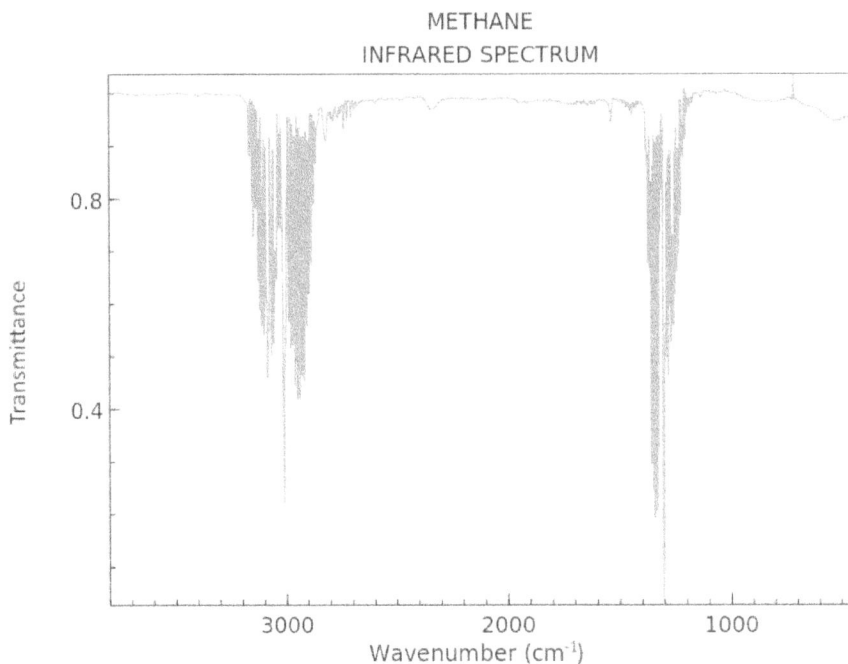

NIST Chemistry WebBook (https://webbook.nist.gov/chemistry)

FIGURE 1.4 Methane absorption spectra with transmittance peaks around 3,300 nm and 7,800 nm. (nm $= 10^7$/wavenumber)

Methane Absorption Spectrum

When the bonds of a methane molecule are excited, they can absorb some of the light energy in a specific wavelength. This characteristic wavelength determines if sensors can detect or not. A sensor's sensitivity depends on the strength of methane absorption at a specific wavelength. In Fig. 1.4 we see a typical methane absorption plot where the transmittance is reduced heavily at specific wavenumber ranges. Understanding these absorption characteristics is key to selecting the right sensor technology for methane detection.

1.4.1 Sensing Types

There are several sensing types associated with measuring methane emissions. These types can range from passive technologies to active technologies or even catalyst type sensors. For the interested readers, please refer to [5] for more details. Here are some brief descriptions of the different types:

- **Passive Sensing** involves an incident light source, typically from the sun, that interacts with the target gas (i.e. methane) and the transmitted light is received into the receptor.

- **Active sensing** involves an actively powered light source that interacts with the target gas before being received into the receptor.

- **Indirect sensing** involves using a substance or target variable, not related directly to the target gas but as a secondary effect that is influenced by the presence of the target gas. An example would be using a CO_2 sensor in Landfill applications to measure methane. This is possible because the makeup of landfill gas is roughly 50% CH_4 and 50% CO_2.

- Other ways of sensing include **chemiresistive sensing** approaches, where a material that is sensitive to changes in the target gas is used. Often times these approaches are not selective to methane exclusively. This means that other gases can cause a change in the measurement leading to false detections or contaiminated signals.

1.4.2   Sensing Modes

Methane measurement using sUAS can be categorized into three primary sensing modes: in situ, column or path-integrated, and imaging-based sensing:

- **In situ sensing** directly measures methane concentration at the sensor's location as the sUAS moves through the air. These sensors, such as tunable diode laser absorption spectroscopy (TDLAS) or electrochemical sensors, provide real-time, localized concentration data with high spatial and temporal resolution. However, in situ measurements require precise flight path planning to intersect the plume effectively and can be sensitive to atmospheric turbulence and local wind variations.

- **Column or path-integrated sensing** measures methane concentration along a defined path between a laser source and a detector, often using open-path laser absorption spectroscopy. This method enables broader spatial coverage and is particularly useful for detecting diffuse sources, but it relies on wind conditions for accurate emission flux calculations and lacks the fine-scale resolution of in situ sensors.

- **Imaging-based sensing** captures methane absorption at specific wavelengths using technologies such as Optical Gas Imaging (OGI) infrared cameras, which provide a visual representation of gas plumes. While imaging is advantageous for rapid leak detection and large-area monitoring, it is less effective for precise quantification and requires specific environmental conditions to optimize sensitivity.

Each of these sensing modes presents unique advantages and challenges, and their selection depends on measurement objectives, emission source characteristics, and operational constraints. In many cases, a hybrid approach combining multiple sensing modes offers the most comprehensive methane emission assessment.

Optical Gas Imaging (OGI)

Optical Gas Imaging (OGI) is a remote sensing technique used to detect and visualize methane and other hydrocarbon emissions using infrared (IR) cameras. OGI cameras operate within specific infrared absorption bands, allowing them to capture gas plumes that are otherwise invisible to the naked eye. A key advantage of OGI is its ability to rapidly scan large areas and identify leaks in real time, making it a preferred tool for Leak Detection and Repair (LDAR) programs. The U.S. Environmental Protection Agency (EPA) recognizes OGI as an approved method for LDAR compliance under 40 CFR Part 60, Subpart OOOOa (commonly referred to as "Quad Oa"). This regulation sets performance-based requirements for OGI surveys, specifying factors such as detection sensitivity, survey frequency, and environmental conditions that affect measurements.

1.4.3 Commonly Used Sensors

A variety of methane sensors are available for airborne measurements, each with distinct advantages and limitations (see Table 1.1). For example, the Los Gatos ultra-portable greenhouse gas analyzer (UGGA, which uses a CRDS) provides exceptional sensitivity (ppb-level), high selectivity, and rapid response times (≈ 1 Hz sampling rate), making it suitable for precise quantification. However, its weight (>5 kg) and power requirements (≈ 70 W) limit its feasibility for sUAS. A more compact alternative is the Axetris Compact LGD (TDLAS), which offers high sensitivity (≈ 100 ppb detection limit) and a fast sample time (≈ 2 Hz) while maintaining a lightweight form factor (≈ 600 g). This makes it a strong candidate for sUAS-based in situ quantification, although its accuracy may be affected by temperature fluctuations and pressure variations.

For path-integrated sensing, the Pergam Laser Falcon (TDLAS) is one of the few commercially available sUAS-compatible open-path sensors. It provides ppm-level sensitivity, real-time response, and a relatively compact design (≈ 1.4 kg), making it well-suited for sUAS-based methane detection. However, its line-of-sight measurement approach means that turbulent conditions or obstacles can impact readings. The DJI U10 (TDLAS-based imaging sensor) improves upon this by providing direct concentration readings with a detection range of ≈ 5–50 m and a sensitivity of ≈ 5 ppm-m, but its accuracy is still highly dependent on atmospheric conditions and viewing angles. In contrast, the Boreal Laser GasFinder3-DC is designed for fixed installations and long-range leak detection, typically mounted on ground-based infrastructure rather than sUASs, due to its larger size and need for precise optical alignment but is very useful for doing fenceline monitoring.

In imaging-based detection, the FLIR GF320 (OGI) offers real-time methane visualization with a detection threshold around 250 ppm-m. This method is particularly useful for qualitative leak identification but lacks precise quantification capabilities. Other solutions, e.g. the Cantronics U-340, can be mounted on sUAS and offer similar levels of detection (if not better). This instrument utilizes a type 2 superlattice

TABLE 1.1 Overview of Commercially Available Methane Sensors for sUAS Applications including cost($=(≤$1,000), $$=($1,000-$10,000), $$$=($10,000-$50,000), $$$$=(≥$50,000)), sensing mode, and sensor type (tunable diode laser absorption spectroscopy (TDLAS), Fourier transformed infrared (FTIR), cavity ring-down spectroscopy (CRDS), optical gas imaging (OGI), quantum cascade laser spectroscopy (QCLS), and metal oxide semiconductor (MOS)).

Sensor Name	Sensor Type	Sensing Mode	Cost	Pros	Limitations
Laser Flacon (Pergam)	TDLAS	Path-integrated	$$$	High sensitivity, effective for remote sensing, suitable for large-area monitoring	Expensive, requires stable environmental conditions, limited in dense or obstructed areas
GasFinder3-DC (Boreal Laser)	TDLAS	Path-integrated	$$$$	Long-range detection, useful for fence-line monitoring	Requires precise alignment, affected by atmospheric conditions
MIRA (Aeris)	FTIR	In situ	$$$	Multi-gas detection, high accuracy	High cost, need heavy-lift sUAS
Hoverguard (ABB)	OA-ICOS	In situ	$$$$	fast response, high accuracy	High cost, need heavy-lift sUAS
SeekIR (SeekOps)	TDLAS	In situ	$$$	High accuracy, lightweight for sUAS integration	Requires direct plume sampling, high power consumption
UGGA (LGR)	CRDS	In situ	$$$	Ultra-sensitive, real-time data, multi-gas capability	Bulky for sUAS integration, high power consumption
U10 (DJI)	TDLAS	Path-integrated	$$$$	Compact, sUAS-compatible, rapid leak detection	Limited quantification capability, sensitive to environmental factors
GF320 (FLIR)	OGI	Imaging	$$$$	Real-time visualization of methane plumes, widely used in industry	Qualitative detection only, affected by temperature contrast and wind conditions
U-340 (Cantronics)	OGI	Imaging	$$$$	Real-time visualization of methane plumes, widely used in industry	Quantitative w/AI, affected by temperature contrast and wind conditions
MIRO	QCLS	In situ	$$$$	High sensitivity, simultaneous detection of multiple gases	Expensive, high power requirement, Not sUAS compatible
TGS2611 (Figaro)	MOS	In situ	$	Low cost, compact, fast response time	Lower sensitivity, prone to drift, affected by humidity and temperature
LGD Compact (Axetris)	TDLAS	In situ	$$	Relatively low cost, compact, quick response time, mid-high sensitivity	requires custom housing and pump

technology to detect the methane at high operating temperatures (HOT, ≈140 K) using cooled filters tuned specifically for detecting methane. The advantage of HOT in OGI cameras, yields longer battery life.

Low-cost alternatives include Metal Oxide Semiconductor (MOS) sensors like the Figaro TGS2611, which provide lightweight integration (≈10 g) but suffer from cross-sensitivity, limited detection range (≈10 ppm), and drift over time.

Each of these sensors presents a trade-off between cost, accuracy, sampling frequency, and environmental robustness. High-precision spectroscopic sensors (e.g., CRDS, QCLS, and TDLAS) enable detailed quantification but are often large, expensive, and power-intensive. Lower-cost alternatives, such as MOS-based sensors, provide lighter and more affordable options but suffer from lower sensitivity, nonlinear response, and long-term drift. For a deeper analysis of methane sensor performance, including comparative evaluations of response times, detection limits, and field deployments, refer to [5], which provides an extensive review of emerging gas sensing technologies.

Pause and Reflect

How do existing methodologies that are adopted in industry skew emissions measurements due to a potential lack of detection in their modes of measurement?

1.5   ON SMART SENSING

When we make typical remote sensing measurements of the environment, we typically aim to extract useful parameter estimates of phenomena we care about. These parameters are then used to draw conclusions about some dynamical process underlying in that phenomena. Over time this process will provide performance feedback, and allow clients or end-users to make adjustments in a similar way to traditional control systems. This loop however is sometimes very slow or limited by the sensors and hardware used to take the measurements and process the data.

With the developments of sensor technologies becoming lighter, cheaper, and often more sensitive, it enables remote sensing with drones to be applied to a wide range of applications. With the developments of the internet of things (IoT) and edge computing, it has become increasingly possible to process data on the edge in real-time. With the developments of data-driven methodologies, and machine learning, it has become increasingly possible to process complex dynamical systems and data on the edge. Furthermore, with the optimal sensor placement and other optimization strategies, edge computing provides a way to, in real-time (or near real-time), give actionable insights to the mobile measurement system – improving data collection by informing where to sense and how to sense. In Chapters 8 and 9, we will discuss more on this topic.

Pause and Reflect

How can the miniaturization of existing sensors in other fields be applied to smart sensing frameworks? How would this transform the way measurements are made?

1.6   CHAPTER SUMMARY

In this chapter we went over the climate change and the global warming problem. We discussed the importance of methane as being a potent greenhouse gas but more importantly why it can and should be used as our 'control knob'. We overviewed the current anthropogenic and biogenic methane sources that contribute to this problem and discussed some policies that are improving the landscape of the measurement and mitigation of methane emissions. We also give an overview of methane sensing types and modes as well as an overview of commonly used methane detection sensors with key sensor characteristics, pros, and limitations listed in a single (Table 3.1).

Bibliography

[1] Landfill Methane Capture — drawdown.org. `https://drawdown.org/solutions/landfill-methane-capture`. [Accessed 21-03-2025].

[2] Methane Might Be a Bigger Climate Problem Than Thought, Study Finds (Published 2022) — nytimes.com. `https://www.nytimes.com/2022/09/29/climate/gas-flaring-climate-methane.html`. [Accessed 21-03-2025].

[3] Tarek Abichou, Nizar Bel Hadj Ali, Sakina Amankwah, Roger Green, and Eric S Howarth. Using ground-and drone-based surface emission monitoring (SEM) data to locate and infer landfill methane emissions. *Methane*, 2(4):440–451, 2023.

[4] Nizar Bel Hadj Ali, Tarek Abichou, and Roger Green. Comparing estimates of fugitive landfill methane emissions using inverse plume modeling obtained with surface emission monitoring (SEM), drone emission monitoring (DEM), and downwind plume emission monitoring (DWPEM). *Journal of the Air & Waste Management Association*, 70(4):410–424, 2020.

[5] Javier Burgués and Santiago Marco. Environmental chemical sensing using small drones: A review. *Science of The Total Environment*, page 141172, 2020.

[6] Derek Hollenbeck, Demitrius Zulevic, and Yangquan Chen. Advanced leak detection and quantification of methane emissions using sUAS. *Drones*, 5(4):117, 2021.

[7] Jacob Mønster, Peter Kjeldsen, and Charlotte Scheutz. Methodologies for measuring fugitive methane emissions from landfills-a review. *Waste Management*, 87:835–859, 2019.

[8] Amy Quinton. Cows and Climate Change — ucdavis.edu. `https://www.ucdavis.edu/food/news/making-cattle-more-sustainable`, Jun 2019. [Accessed 13-04-2025].

[9] JR Roscioli, TI Yacovitch, C Floerchinger, AL Mitchell, DS Tkacik, R Subramanian, DM Martinez, TL Vaughn, L Williams, D Zimmerle, et al. Measurements of methane emissions from natural gas gathering facilities and processing plants: measurement methods. *Atmospheric Measurement Techniques*, 8(5):2017–2035, 2015.

[10] Drew Shindell and Christopher J Smith. Climate and air-quality benefits of a realistic phase-out of fossil fuels. *Nature*, 573(7774):408–411, 2019.

CHAPTER 2

Emission Source Detection

A s we have learned in the previous chapter, methane has several modes of sensing and depending on the mode of choice, there are several technologies to provide measurements (each with their limitations). In this chapter, we will explore how those limitations impact the detection process, as well as how the deployment of the sUAS can affect it in general.

2.1 sUAS PLATFORMS AND SENSOR INTEGRATION

There are two key factors that enable the detection of methane onboard sUAS. The first is the type of sUAS platform used for sensor integration. The second lies within the sensor integration process itself. Depending on the specific application of interest, the choice of sUAS platform and how the sensor is integrated will be unique.

2.1.1 sUAS Platforms

In the past decade, sUAS have been used in a host of research applications to learn about our environment or to improve a task, or in developing more robust sUAS systems in general. In this book, we will categorize them into three generalized categories: fixed-wing, multirotor (or copter), and hybrid vertical-takeoff-and-landing (VTOL). Examples of each type of sUAS can be seen in Fig. 2.1. Although, it is not covered here, there is another class of sUAS called lighter-than-air (LTA) that can be used to provide long-term monitoring. The LTA type uses lifting gases, such as helium or hydrogen to provide a buoyancy force and is typically more sensitive to wind gusts and drift. **Multirotor** platforms are sUAS that include two or more propellers. The platform uses these propellers as the main source of lift. The flight control is carried out by adjusting the speed of each rotor to induce a pitch, roll, or yaw movement (see more details regarding vehicle dynamics in [4]). These platforms are particularly good in conducting aerial photography, surveillance, mapping, or other functions requiring precision and full control in the spatial position. A common reference to drones typically refers to this type of sUAS. **Fixed-wing** platforms are sUAS comprising of rigid wings to fly like a traditional manned aircraft plane. The primary means of lift is gained from the airfoil design of the wings and is exceptionally more efficient than multirotor platform. The flight control is undertaken by means of an alerion, elevator,

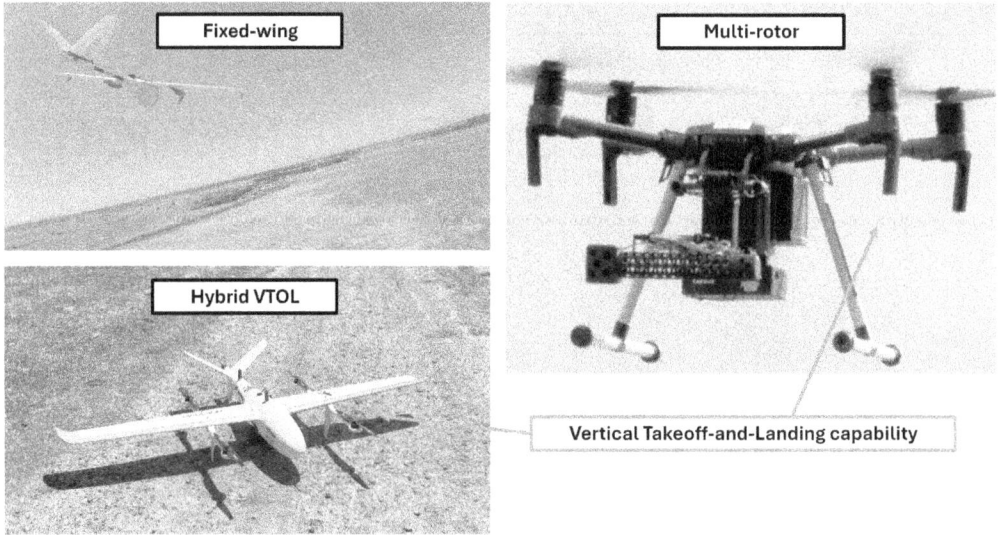

FIGURE 2.1 Examples of the common sUAS platform types.

and rudder command which conversely induces roll, pitch, or yaw movement, respectively. The flight dynamics of fixed-wings are not able to achieve the same position control seen with multirotor platforms and as such, also can make fixed-wings good for covering large areas in mapping/remote sensing applications. More information on flight dynamics of fixed-wings can be seen in [2]. **Hybrid VTOL** platforms are a combination of multirotor and fixed-wing attributes. Typically, these platforms are designed as a traditional fixed-wing platform and are modified to include a three-or four-rotor design for VTOL. The flight control can be done in VTOL mode (in the same manner as a multirotor) or in fixed-wing mode. This functionality gives more operational flexibility considering that large open spaces are required for fixed-wing platforms to land and take off (usually done with landing gears or in a belly landing fashion).

2.1.2 Sensor Integration

Sensor integration is the process by which a physical sensor is fixed to the sUAS. This process also includes how the sensors data is retrieved, processed, and or transferred to the ground control station (referred to as the data acquisition system). Many times this process can be case-specific to the unique combination of sensor and platform. In this section, we will focus on the generalities of sensor placement and reserve the specifics associated with the data acquisition system to be left to the interested practitioner. The placement of methane sensors on a small unmanned aerial system (sUAS) significantly influences their ability to detect and quantify emissions. Different mounting configurations (see Fig. 2.2) – such as joust or boom-mounted, top-mounted, and bottom-mounted – each has advantages and challenges that must be carefully considered based on the sensor type, sensing mode, and mission objectives. Additionally,

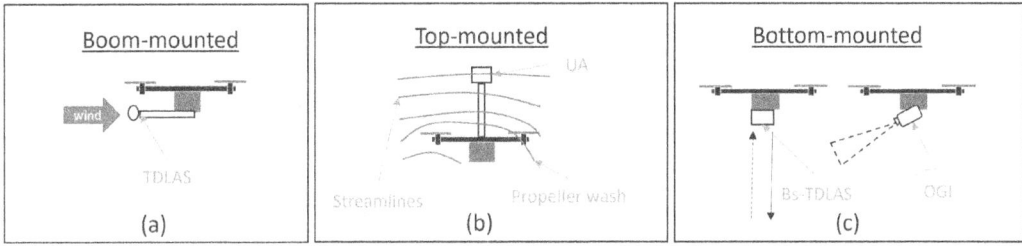

FIGURE 2.2   Sensor integration examples of (a) TDLAS boom-mounted, (b) ultrasonic anemometer (UA) top-mounted, and (c) a bs-TDLAS and OGI in bottom-mounted configurations.

factors like the 'propeller wash effect', electromagnetic interference (EMI), structural material (e.g. carbon fiber) influences, and shielding requirements further complicate integration choices.

Joust or Boom-Mounted Sensors – A joust or boom-mounted configuration extends the sensor away from the main body of the drone, typically in front of or below the platform. This setup minimizes wake turbulence and avoids sampling air that has been disturbed by the drone's propellers. For in situ gas analyzers–such as cavity ring-down spectroscopy (CRDS) or tunable diode laser absorption spectroscopy (TDLAS) – this positioning is crucial, as these instruments rely on sampling undisturbed ambient air. If placed too close to the propellers, the induced mixing may dilute the methane plume, reducing the measured concentration. Conversely, in some cases, increased mixing may enhance detection by broadening the plume's extent, allowing for a larger sampling window. However, boom-mounted sensors increase aerodynamic drag and may affect flight stability, especially in turbulent conditions. Additionally, long booms may introduce mechanical vibrations, which can impact sensitive optical-based instruments, necessitating vibration-damping mounts.

Top-Mounted Sensors – Top-mounted configurations are less common for methane detection but may be advantageous for imaging-based sensors, such as optical gas imaging (OGI) cameras, which rely on a downward or forward-facing perspective. However, in situ gas sensors placed in a top-mounted position are generally less effective, as methane plumes tend to rise due to buoyancy and atmospheric mixing, making it difficult to capture high-concentration air samples. Additionally, many drones place antennas and telemetry systems on the top, increasing the likelihood of EMI interference that could affect sensor electronics, particularly laser-based or spectroscopic measurement systems.

Bottom-Mounted Sensors – Bottom-mounted sensors are the most common configuration for path-integrated and in situ measurement systems. Many TDLAS-based systems, such as the DJI U10, use this configuration to provide an unobstructed, downward-facing measurement path, reducing the likelihood of interference from drone components. Path-integrated sensors, which detect methane along an optical path between an emitter and detector, require a clear field of view and benefit from this mounting approach. In situ sensors also benefit from a bottom-mounted configuration since methane emissions often remain closer to the ground, particularly in low-wind conditions.

However, propeller downwash presents a key challenge for bottom-mounted sensors. The downward air currents generated by the rotors can dilute or displace methane plumes, significantly altering the measured concentration depending on wind conditions and drone altitude. In some cases, strong downwash may push methane away from the sampling region before detection occurs. This issue is especially relevant for fixed-position hovering measurements, where the ground effect can further amplify disturbances. For mobile measurements, a horizontal flight path with the sensor positioned away from the drone's main downwash region can mitigate these effects. Additionally, placing the sensor on an extended boom outside of the propeller wash can help isolate it from the disturbed airflow [6]. Furthermore, once the sensor is outside of this disturbance zone and if the wind speed (or relative wind speed) over the sensor is above 2 ms^{-1} then a clear air sample can be adequately measured.

Electromagnetic and Structural Considerations – Beyond physical placement, the choice of drone materials and electronic shielding plays a crucial role in maintaining sensor accuracy. Carbon fiber frames, commonly used in high-performance drones, can create electromagnetic shielding effects, potentially interfering with radio-based sensors or open-path laser spectroscopy that requires precise alignment. Moreover, telemetry signals, onboard Wi-Fi, and radio frequency (RF) communication systems can introduce unwanted noise into high-precision spectrometers and mass analyzers, or similar techniques. Proper grounding and electrical shielding using conductive materials or shielding enclosures is essential to prevent signal corruption.

Comparison Across Sensor Types and Modes – Different sensor types and sensing modes impose distinct constraints on mounting configurations. In situ sensors typically require exposure to undisturbed air, making boom-mounted configurations preferable to avoid wake and propeller wash effects. However, if in situ sensors are placed in a bottom-mounted configuration and utilize a pneumatic pump system, they can avoid the same effects. In that case, the pneumatic tubing can be extended in a boom-mounted way or draped a large distance from the aircraft, such as those seen in [1]. Path-integrated sensors, such as TDLAS and Fourier transform infrared (FTIR) systems, benefit from bottom mounting to maintain an unobstructed optical path between the sUAS and the ground. Imaging-based sensors, including OGI cameras, may be top- or forward-mounted, depending on the desired perspective for leak visualization. Ultimately, sensor placement must balance aerodynamic effects, environmental exposure, and interference mitigation to optimize methane detection performance.

Now let us turn our attention to the fixed-wing aircraft. This type of sUAS has many desirable characteristics (e.g. long endurance, large spatial coverage, etc.) as well as some drawbacks (e.g. no stopping). The principles of operations require the aircraft to maintain an airspeed above the stall speed, at which the airflow over the wing separates and the lifting force ceases to be useful (i.e. gravity starts to win). This also means that sensors cannot be placed within certain envelopes of critical flight components (e.g. disrupting airflow over the wing, control surfaces, or near propulsion systems). Understanding the aerodynamics around the aircraft provide insight into key locations for potential sensor integrations that depend on airflow. Some common examples are seen underneath the wing (e.g. near or on the pylon) or at the nose of the aircraft (e.g. collocated with the pitot-static tube). This makes in situ gas sampling more challenging, as the air intake and sampling rate must be designed to handle high-speed airflow. Sensors such as cavity ring-down spectroscopy (CRDS) or tunable diode laser absorption spectroscopy (TDLAS) require careful placement to avoid wake turbulence and ensure that sampled air is representative of the plume. For sensors that do not necessarily need clean airflow into the sensing region (e.g. backscatter TDLAS), the integration can be placed into the belly of the aircraft, often protected by an actuated door or sliding cover. Fig. 2.3 illustrates some potential integration locations.

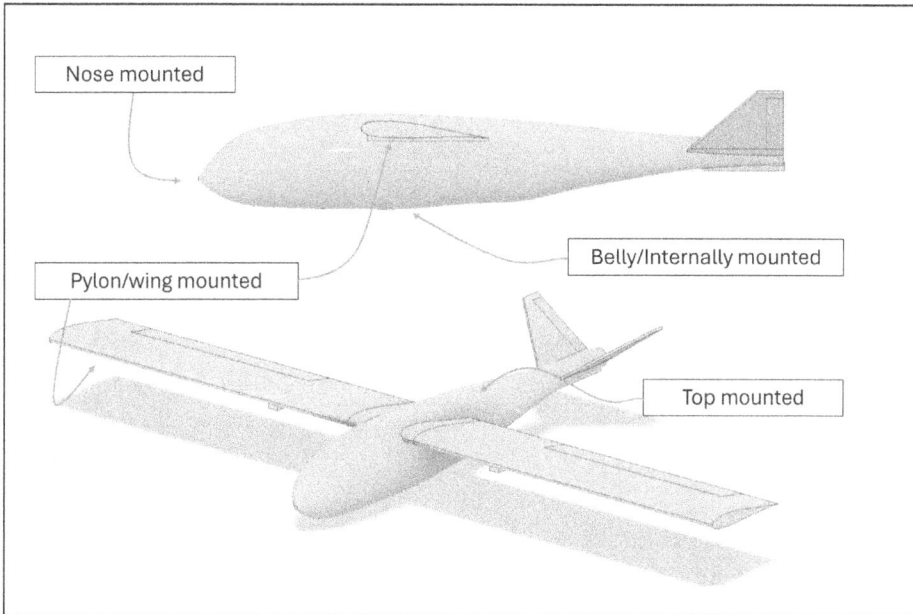

FIGURE 2.3 Potential locations to consider for sensor integration on fixed-wing sUAS.

2.2 MOBILE SENSOR DEPLOYMENT

Deployment – is defined here as the movement of the mobile sensor system along a specified trajectory in order to achieve an objective or a series of objectives. Deployment of the sUAS is the next critical step toward reaching the detection goal of atmospheric trace gases. In later chapters, we will see how deployment can influence the performance of localization and quantification of emission sources. One of the key factors in detection success is understanding how the environmental conditions impact the sensors ability to detect (e.g. wind speed, atmospheric turbulence, humidity, barometric pressure, temperature, etc.). Additionally, within the flight area, the sUAS operator needs to understand the location of likely emission sources, the current wind/atmospheric conditions, and the sensor characteristics (e.g. sensing mode, rise/relaxation times, accuracy, precision, sensitivity, saturation level, etc.). The likely emission sources (also called a priori) are often known before the flight campaigns are conducted. Given the sensing mode, a different deployment trajectory may be executed. Examples of different sensing modes for SEM can be seen in Fig. 2.4. By utilizing the Gaussian plume model as a guide, for example, data aware decisions on path planning can be used to define deployment trajectories to ensure good detection probabilities [5].

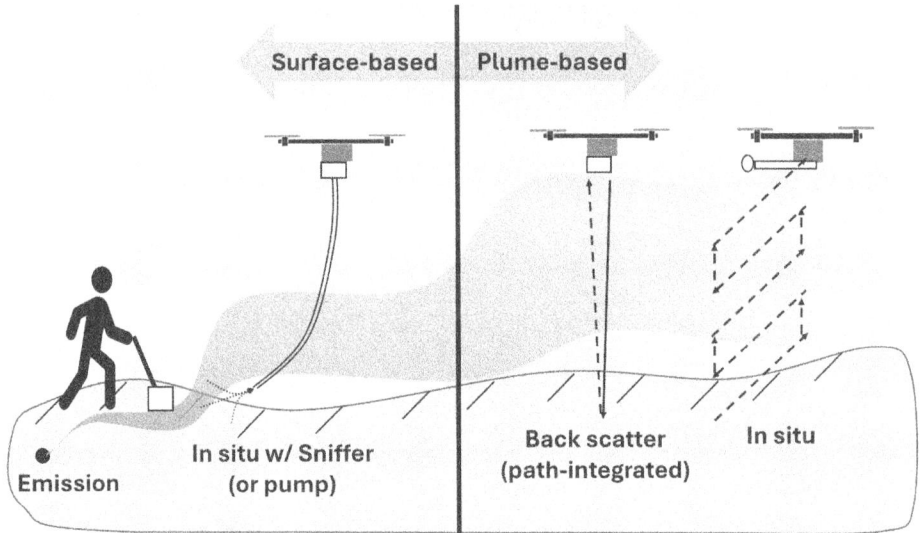

FIGURE 2.4   A diagram comparing between surface-based and plume-based detection using multirotor platforms.

The specific trajectory for the deployment will depend on the sensor type, sensing mode, integration configuration, and the application. For example, if we look at the application of landfills with area sources, in situ sensors can be applied in different ways. As seen in Fig. 2.4, in the case of pump-based in situ TDLAS, the system can be applied in two major ways. The first is to connect a long pneumatic tube to measure flux at the surface. The second is to connect the pneumatic tubing in a boom-mounted configuration to measure in the plume, The first configuration is more ideal for surface measurements, making them more robust to wind conditions and reduce the potential disturbances from the aircraft itself (e.g. propeller down wash). The second configuration, if applied in the same way, may suffer detection-related issues. However, if the second configuration is deployed in a plume-based approach (focusing more on capturing the advected plume at some distance downwind), the detection capabilities would improve. While in this example, it is possible for the first configuration to be applied with the in-plume approach, it may need extra compensation to accurately place the sensing region in the correct location. This would also introduce some uncertainty due to fluctuations in the tube position as the sUAS moves throughout the survey area. Therefore, it becomes clear that the choice of deployment strategy depends heavily on the combination of sensor type, mode, configuration, and application. Once the deployment strategy is aligned with the sensor integration step, the concentration measurements in space can be collected and used to infer/estimate the location of the source or make quantification estimates of the emission source rate (see Chapters 3 and 4).

2.3 PROBABILITY AND MINIMUM DETECTION LEVEL

The deployment and the sensor integration strategy is not the only factor that affects the detection of methane in the atmosphere. Given variability in the source type or in the weather conditions, detection can become a daunting task. The question then arises, *how can we know we have surveyed enough?* If we have a continuous emission source, *what are the chances of detecting that emission source given the sensor integration, the sensor characteristics and how the sUAS is deployed?* However, if we have an emission source that is intermittent, the chances of detecting the leak depend heavily on the intermittent properties of the source (e.g. are they weather depended?, does it have a specific frequency or duration?, etc.). The concept of probability of detection can be introduced to help put this issue into perspective from an operational standpoint. Let us consider for a moment that if an emission source is present only a fraction of the time, how many times must we survey it to be at least 80% confident that there is no leak? This kind of question is reflected in Fig. 2.5 as contour lines (each line represents a true probability of detection of 80%). Additionally we can include generalized sensor characterizations, such as a detection probability, associated with the sensitivity and working principle of the sensor itself. This means that if the sensor is indeed within detectable range of the emission concentration but does not detect it, a probability is assigned for how often it does. When you combine these two ideas you get the curves in Fig. 2.5.

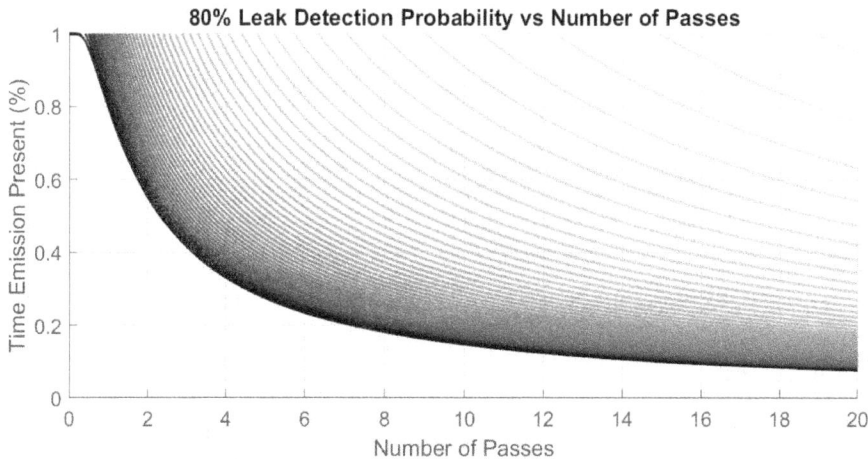

FIGURE 2.5 This diagram illustrates the behavior of the probability of detection given an intermittent source as a function of a sensor dependent detection constant. The darkest lines indicate a perfect sensor with a detection constant of 1. As the lines get lighter, the detection constant goes from 1 to 0.

For example, if we had an emission leak present only 50% of the time, we can observe from Fig. 2.5 that in order to get a probability of detection of at least 80% or greater, we would need to have at least two visits (potentially three) with a perfect sensor ($P_d = 1$), and we would need on the order of 15 visits with a poor performing sensor ($P_d = 0.2$). This kind of conversation has been happening more and more within the research and industrial community, see [3]. The authors conducted extensive controlled release experiments, both fully and semi-blinded, to derive continuous probability of detection (POD) functions for each technology (Bridger Photonics Inc.'s Gas Mapping Lidar, Kairos LeakSurveyor, and NASA/JPL's Airborne Visible/Infrared Imagine Spectrometer-Next Generation (AVIRIS-NG)). These functions enable the calculation of detection probabilities based on variables such as source emission rate, wind speed, and flight altitude. The probability of detection (POD) formulation in [3] relies on,

$$POD(x) = F(g(x; \phi); \theta), \tag{2.1}$$

where the function $g(\cdot)$, represents a predictor function, and $F(\cdot)$, represents a continuous inverse link function. The coefficients ϕ and θ, in this context, can be solved using a maximum likelihood approach with the Bernoulli distribution. The interested reader should consult [3] for a generalized predictor model and candidate inverse link functions. Additionally, the study developed quantification uncertainty models that address measurement bias, variability, and precision, allowing for direct calculation of the true source rate distribution from the estimated measurements. The findings demonstrate the potential of all three technologies that the authors explored in methane detection and mitigation efforts. The methodologies outlined can be applied to other measurement techniques or updated with future controlled release experiments. The study also emphasizes the importance of using data from a range of sites and times to avoid underestimating measurement uncertainties.

Pause and Reflect

How can automation of mobile sensor systems be deployed to detect emissions when weather conditions are optimal? How could this scale our mitigation efforts, improve detection, and reduce uncertainty?

2.4   CHAPTER SUMMARY

In this chapter we discussed the three major types of sUAS (multi-rotor, fixed-wing, and hybrid VTOL). Then, we compared the differences in sensor integration considerations depending on which platform is chosen, as well as with which sensor type, sensing mode, and integration configuration is used. Depending on this combination, the deployment strategy will be designed to maximize the signal detection of the emission source. Although, based on the sensor characteristics and integration, the measurement system may have a detection factor associated with it that needs to be taken into consideration. Furthermore, weather and other conditions need to be assessed for their contribution to the overall probability of detection and uncertainty quantification methods.

Bibliography

[1] Tarek Abichou, Nizar Bel Hadj Ali, Sakina Amankwah, Roger Green, and Eric S Howarth. Using ground-and drone-based surface emission monitoring (SEM) data to locate and infer landfill methane emissions. *Methane*, 2(4):440–451, 2023.

[2] Randal W Beard and Timothy W McLain. *Small Unmanned Aircraft: Theory and Practice*. Princeton university press, 2012.

[3] Bradley M Conrad, David R Tyner, and Matthew R Johnson. Robust probabilities of detection and quantification uncertainty for aerial methane detection: Examples for three airborne technologies. *Remote Sensing of Environment*, 288:113499, 2023.

[4] Pulin K Garg. *Unmanned Aerial Vehicles: An Introduction*. Mercury Learning and Information, 2021.

[5] Derek Hollenbeck, Moataz Dahabra, Lance E Christensen, and YangQuan Chen. Data quality aware flight mission design for fugitive methane sniffing using fixed wing sUAS. In *Proc. of the 2019 International Conference on Unmanned Aircraft Systems (ICUAS)*, pages 813–818. IEEE, 2019.

[6] Brendan Smith, Garrett John, Brandon Stark, Lance E Christensen, and YangQuan Chen. Applicability of unmanned aerial systems for leak detection. In *Proc. of the 2016 International Conference on Unmanned Aircraft Systems (ICUAS)*, pages 1220–1227. IEEE, 2016.

CHAPTER 3

Emission Source Localization

G Iven that a sensor has been integrated onto a sUAS platform and that system can be deployed successfully, in the detection sense, what can we do? Two of the common questions asked are: where is the source located? (i.e. source localization), and how much is being emitted? (i.e. emission quantification). The goal of this chapter will be to address the localization, by defining the general problem of localizing a source, going over some localization methods used (e.g. heuristics, and model-based) and then discussing some open challenges. Some of the methods involve aspects of quantification and will be a focus of Chapter 4.

3.1 THE SOURCE LOCALIZATION PROBLEM

The source localization problem can be defined in several ways, depending on which literature you read. The general goal of **localization** is to determine the spatial position of the emission source (or sources) based on a series of sensor observations (typically concentration and wind data). There are three main approaches to the localization problem: direct search methods, grid-based methods, and inverse dispersion/inference-based methods.

- A **direct search** (referred to as source seeking or chemical plume tracing) strategy actively moves the mobile sensing system (e.g. drone or robotic platform) toward the emission source without creating a full spatial model.

- The **grid-based search** methods use the spatial concentration information to generate a map over the area of interest, which can be analyzed in post processing, typically by scanning the entire area of interest.

- The **inference-based** methods estimate the source location mathematically through dispersion models and/or statistical inference techniques rather than direct search or mapping approach (e.g. like those found in grid-based search).

A diagram depicting these localization approaches is illustrated in Fig. 3.1.

DOI: 10.1201/9781003669470-3

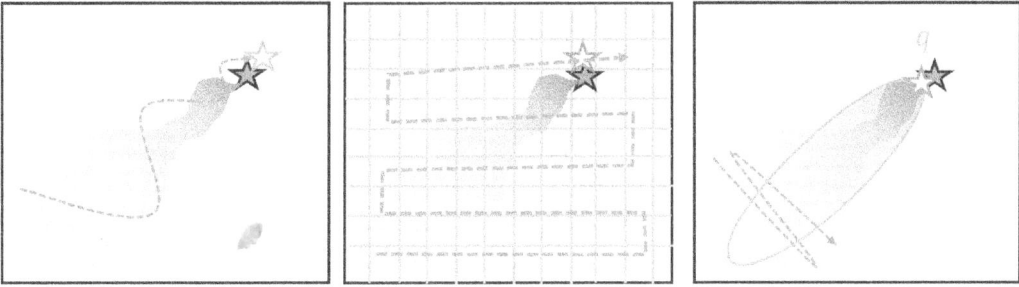

FIGURE 3.1 Some different types of localization are represented in the diagram. (left) source seeking, (middle) grid-based, and (right) inverse dispersion and inference.

3.2 LOCALIZATION METHODS

3.2.1 Direct Search

In the source seeking world, for example, heuristic and gradient-based methods are used to trace the plume back to the source. Some notable heuristic methods include casting, spiral surge, and anemotaxis. Variations of these heuristic methods include bio-inspired versions (e.g. zig-zagging moth-inspired). Other methods directly rely on the gradient information (e.g. like with chemotaxis) or with probabilistic models, thereby maximizing information in some sense (e.g. infotaxis [25], and entrotaxis [12]). One prominent category includes bio-inspired algorithms like casting and anemotaxis, which mimic insect behavior by iteratively moving upwind toward higher concentrations. These methods are effective in real-world conditions with turbulent dispersion but may struggle in low-wind environments where concentration gradients are weak, noisy or misleading. Some methods use the maximum likelihood estimator combined with the gradient of the Cramer Rao lower bound (CRLB) [17], aiming to use the gradient information to reduce the estimation uncertainty. Alternatively, physics-based approaches can be used that take advantage of inherent properties and theorems, such as the divergence theorem in the original fluxotaxis algorithm [27] or the fractional calculus-based fluxotaxis [9]. The advantages of direct search algorithms include their ability to continuously adjust their path based on live sensor data. Unlike model-based approaches, direct search does not require predefined gas dispersion models. When wind conditions are stable and the source is sufficiently strong, direct search can efficiently localize emissions. Making them ideal for dynamic or complex terrains where predefined gas distribution models are expensively impractical. However, direct search methods do have their drawbacks. For example, wind fluctuations, turbulence, and low sensor sensitivity can lead to misguidance. In low-concentration conditions, small gradients may not provide clear directional cues. They can get trapped in areas of temporary high concentration without reaching the true source. If the search space is vast, exploration can become inefficient compared to model-driven methods.

3.2.2 Grid-based Methods

Examples of grid-based search include occupancy grid mapping [21], Gaussian process regression [20], or kriging interpolation [26]. Another approach utilizes the Gaussian kernel to estimate the spatial concentration map [13]. Each method has some benefits and limitations. For example, occupancy grid mapping may require a dense sensor network (distributed spatially across the domain of interest), or sufficient spatial coverage from a mobile sensor over a reasonable time-scale to construct the map. Gaussian process regression may be an improvement to this approach but increases the computational complexity. Other methods may suffer with accuracy from turbulent conditions or even violate the implied assumptions of the original method (e.g. with ordinary/simple kriging – we will see more in Chapter 4).

Grid-based methods discretize the environment into uniform cells, enabling spatially organized data collection and provides a clear gas concentration distribution across the entire search area, making it useful for large-scale monitoring. Unlike direct search methods, grid-based approaches are not as sensitive to transient wind fluctuations. It works well for continuous monitoring of emissions over time and space. However, as grid resolution increases, data processing and storage requirements grow significantly which requires systematic area coverage, making it slower than direct search methods in certain conditions. High spatial resolution requires a large number of sensor trajectory measurements, which may be impractical in some field deployments. Static grid-based approaches may struggle with highly dynamic sources or rapidly changing environmental conditions.

3.2.3 Inverse Dispersion and Inference Methods

Examples of inverse dispersion and statistical inference methods include the Gaussian plume model inversion [24, 22, 10, 7, 8]. In some cases the inference methods and source seeking are combined to achieve an improved results, like in the work of [11] where a particle filter and Gaussian plume model are used to sequentially update the source term parameters in the Bayesian sense. Other methods that incorporate computational fluid dynamic (CFD) models, can simulate plume dispersion and can conduct optimization for solving the localization problem but suffer from high computational cost. CFD-based methods are also well suited for accuracy in complex environments despite their drawbacks. Lagrangian-based modeling (e.g. Stochastic Time-Inverted Lagrangian Transport – STILT [14]) has shown promise in localizing sources, but the spatial scale at which they are applied, is typically large. Sparse sensor placement is another class of methods that can improve the way we look at inference methods. For example, in the work of [15], the sparse set of sensor locations are chosen such that they maximize the ability to reconstruct the flow field. This work uses ideas of compressed sensing [1] and proper orthogonal decomposition. This idea is beneficial to our problem as it looks for the areas that are most informative to use a an anchor point for the inversion model (we will see more on this in Chapters 8 and 9).

An illustrative example of source localization can be seen from the Lambert W function. Following the work by Matthes et al (2005) [16], Carslaw (1959 [2], and

Roberts (1923) [19], the solution to a single point source advection diffusion equation (ADE) can be solved for a dynamic system approximately by making a quasi-steady state assumption if the variance and transient behavior of the wind small. The ADE equation is shown here (written in Einstein notation),

$$\frac{\partial y}{\partial t} - D\frac{\partial^2 y}{\partial x_i^2} + v\frac{\partial y}{\partial x_i} = q_0\delta(t - t_0)\delta(x_i - x_{si}), \tag{3.1}$$

where y is the concentration, D is the diffusion coefficient, q_0 is the source rate, and i represents the individual spatial directions, e.g. $\mathbf{x} = [x_1, x_2, x_3]^T$. The source location is given as $\mathbf{x}_s$. The Lambert function is denoted as W_0 and has the property,

$$W_0 e^{W_0} = z. \tag{3.2}$$

The solution of (3.1) can be solved for concentration as a function of distance, d,

$$\bar{y}(\bar{x}_1, \mathbf{x}_s, q_0)_i = \frac{q_0 \exp(\frac{\bar{u}(\bar{x}_1 - x_{s1})}{2D})}{\pi^{\frac{2}{3}} D d}. \tag{3.3}$$

Using the Lambert function, one can rewrite the formulation to output distance given the concentration, initial guess of the source location, and the average lateral distance from the source,

$$d_i(y_i, \bar{x}_1, \mathbf{x}_s, q_0) \approx \frac{2D}{\bar{u}} W_0(\frac{\bar{u}q_0}{4\pi D^2 y_i} \exp(\frac{\bar{u}}{2D}(\bar{x}_1 - x_{s1}))). \tag{3.4}$$

The distance function is in the form of a level set and can be used to solve for, $\bar{x}_2$,

$$x_{s2,i} = \bar{x}_{2,i} \pm \sqrt{d_i^2 - (\bar{x}_1 - x_{s1,i})^2}. \tag{3.5}$$

Then by using a cost function of the form,

$$\min_{q_0, x_{2,0}} : \sum_{i,j=1}^{m} (x_{1,i}(x_{2,0}, q_0) - x_{1,j}(x_{2,0}, q_0))^2, \tag{3.6}$$

an estimate of both q_0 and $\hat{\mathbf{x}}_s$ can be made. Although, this chapter we are primarily looking at localization, it is important to note that some localization methods can offer a way to quantify as well.

As a bonus, what if we wanted to create an autonomous survey that is capable of sampling the plume and making this localization and quantification estimate using Lambert W function? Well, one approach would be to conduct a random search along a constrained flight path. In the foraging literature, the Lévy walk has been shown to be effective at searching sparse environments. However, Brownian motion is more efficient in dense areas. This adaptive search model [18] can switch dynamically from Lévy to Brownian (see Fig. 3.2) based on observing concentration signals above a specified threshold. The switching mechanism is using a tumble probability $P(z(t))$, such that $z(t)$ is governed by the stochastic differential equation (SDE) below,

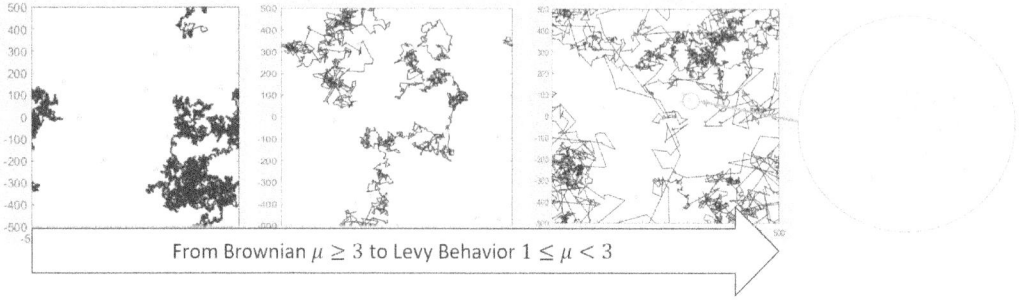

From Brownian $\mu \geq 3$ to Levy Behavior $1 \leq \mu < 3$

FIGURE 3.2 This diagram depicts the change in searching strategy from Brownian motion to Lévy following the work in [18].

$$P(z(t)) = e^{-z(t)}, \quad 0 \leq z \leq 5 \tag{3.7}$$

$$\dot{z} = -\frac{\partial U}{\partial z}A + \epsilon, \begin{cases} U = (z-h)^2, \epsilon : \begin{cases} H = \frac{1}{2}, \mathrm{N}(0, \sigma) \\ H \neq \frac{1}{2}, \mathrm{fGn} \end{cases} \\ A = \max(A_{min}, \alpha(t)) \end{cases} \tag{3.8}$$

$$\alpha_k = C_\alpha \alpha_{k-1} + k_t F \begin{cases} F = 1, \text{found target} \\ F = 0, \text{otherwise} \end{cases}, \tag{3.9}$$

where $k_t > 0$ is the gain and U is the potential function. We extend [18] by adding, fractional Gaussian noise (fGn) to the SDE. Fractional Gaussian Noise (fGn) can be derived in a similar manner as Gaussian noise (Gn). For example, by taking the difference between successive steps in Brownian motion (Bm), one can arrive at Gn. The Riemann-Liouville fractional integral can be used to define fractional Brownian motion (fBm) and is given below,

$$B_H(t) = \frac{1}{\Gamma(H + 1/2)} \int_0^t (t-s)^{H-1/2} dB(s). \tag{3.10}$$

Here the term $dB(s)$ is the general definition of white noise, the term H is the Hurst parameter and $\Gamma(\cdot)$ is the usual gamma function. We can see that depending on the Hurst parameter value (between 0 and 1) the motion can be

- Brownian motion with $H = 1/2$

- Positively correlated $H > 1/2$

- Negatively correlated $H < 1/2$.

For $H > 0.5$ the process exhibits long-range dependence such that,

$$\sum_{n=1}^{\infty} E[B_H(1)(B_H(n+1) - B_H(n))] = \infty. \tag{3.11}$$

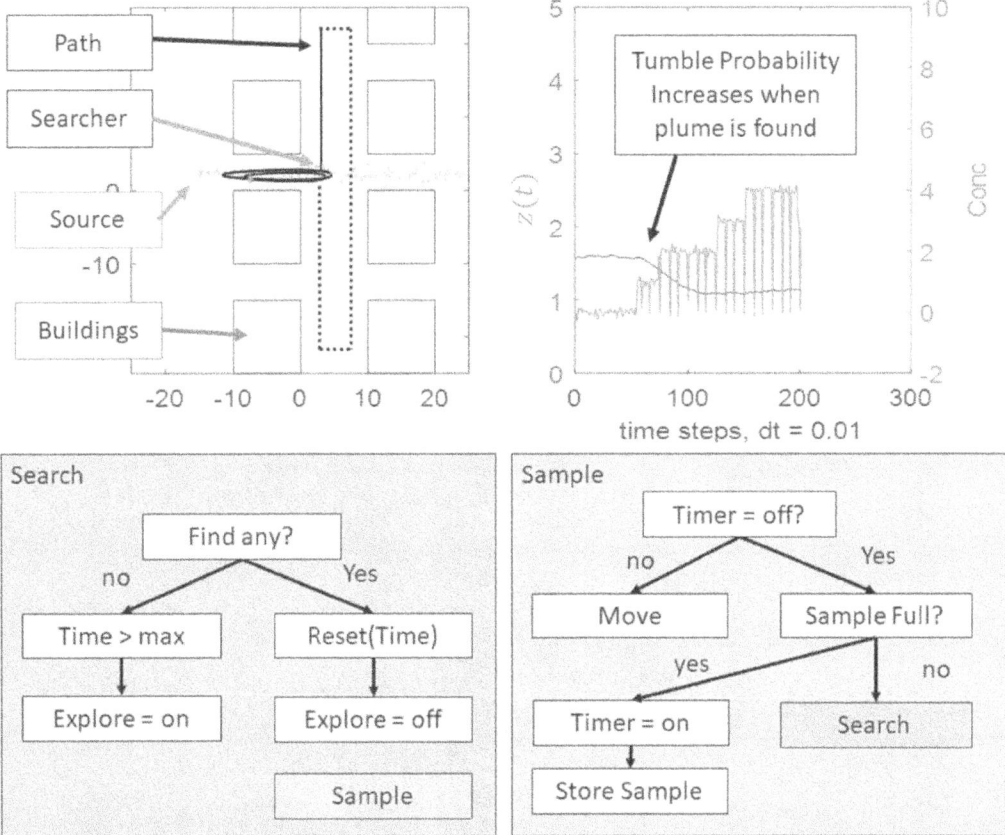

FIGURE 3.3 (top) Depicts the autonomous searching path where the searcher detects the emission and switches the searching strategy to a Brownian-based random walk, slowing transitioning to a Lévy walk before continuing on the path. (bottom) Schematic of the switching algorithm.

We can denote the fGn based on Hurst parameter as, $fGn_H(k) = B_H(k + 1) - B_H(k)$ where k here is used as a discrete time step and not the gain used in the control diagram. In our case we use low pass filtered white noise with high cutoff ratio initially. This process is sometimes referred to as an Ornstein-Uhlenbeck (OU) process. Replacing white noise with fGn creates a fractional OU process with new tuning parameter H. To calculate or compute the fractional integral one can use the Gaussian quadrature, Cholesky decomposition method, or circulant embedding by [5]. This adaptive search model can adjust from Brownian motion to Levy walks in a 2D random search. By reducing the problem to a 1D path problem (i.e. survey route) adding decision trees and modeling fugitive gas with a small time-scale filament model [6] (more on this in Chapter 6) we have the opportunity to optimize the random search for application (see Fig. 3.3). To do this we need to gather enough information to form a set of sample data sets to use in the inversion method for a zeroth order approximation of source localization ($\hat{\mathbf{x}}_s$) and quantification (q_0).

Using the quasi-steady inversion method on experimental data, we can see the results from just two samples in the presence of two sources (circles) in Fig. 3.4, reveal

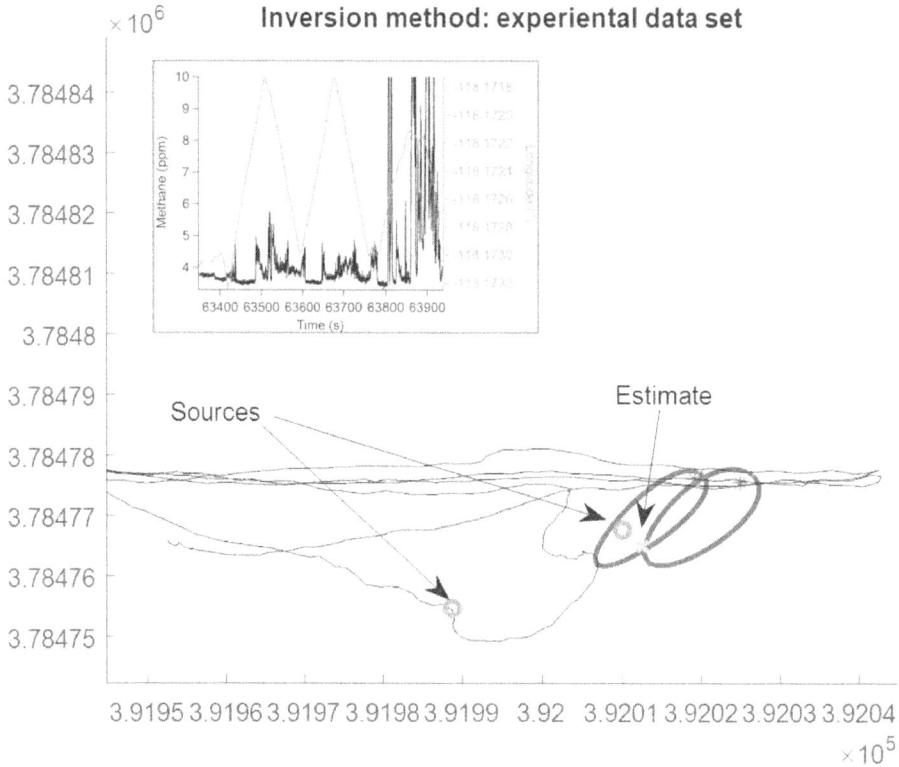

FIGURE 3.4 An experimental example applying the quasi-steady state inversion method to localize a real-world leak. In this case, two sources are present but based on the measured data (shown in the reduced image) only one of the sources can be located.

only one solution. Only taking a small section of raw data from each longitudinal pass we can approximate the source location (x) from our measurement with the open path laser spectrometer (OPLS) [3]. It should be noted, that in the presence of multiple leaks, this methodology still has limitations and requires further investigation.

To recap, the inverse dispersion and inference methods discussed in this section leverage statistical inference, Bayesian estimation or inverse dispersion modeling to infer source locations with a strong theoretical foundation. By incorporating environmental factors such as wind speed, turbulence, and atmospheric stability to refine predictions, one can infer source locations even with a limited number of sensors by solving inverse problems (which are often ill-posed – see Chapter 8). Many inference techniques offer confidence intervals or probability distributions for estimated source locations, making them more ideal in some cases where systems are noisy. However, performance of the methods discussed depends on the accuracy of the wind dispersion models and the general errors in modeling, which can lead to imprecise or incorrect localizations. Some inference methods, such as Bayesian approaches, can be computationally expensive (depending on the type of modeling used), making real-time applications challenging. Uncertainties in sensor readings can propagate through the inverse modeling process and ultimately affecting localization accuracy as well. Thus these methods generally work best for stationary or semi-stationary sources – meaning, rapidly changing emissions may reduce reliability.

3.2.4 Control Volume Approach

In practical applications, many emission sources manifest as point sources from individual pieces of equipment (e.g. well heads, tanks, etc.) or other components (e.g. valves or hatches). This equipment or infrastructure is typically constrained on individualized pads that maintain a specified distance from each other. In this scenario, we can look at each pad individually until a detection event is observed. To find out if the detection occurred from a specific site (i.e. leaking or not), we need to understand the flux entering or leaving control volume surrounding the site (e.g. an imaginary box). We know that if a site is not leaking, it will produce a net zero flux on each surface of the control volume. This can be seen in the surface integral used in Gauss's theorem,

$$\iiint_V (\nabla \cdot c\mathbf{u})dV = \oiint_S (c\mathbf{u} \cdot \hat{n})dS, \tag{3.12}$$

where the normal vector, $\hat{n}$, points outward perpendicular to the surface. When aligned with the wind vector, $\mathbf{u}$, the dot product produces a positive value and a negative value when opposed. The multiplication between the scalar concentration values and the wind vector is referred to as the flux. The Gauss Divergence Theorem states that the total flux of a vector field (such as the gas concentration flux) through a closed surface is equal to the integral of the divergence of that field within the volume (equation (3.12)). In this application, if the net flux entering and leaving the defined perimeter is non-zero (i.e. greater than some threshold), a gas source exists inside the control volume. Alternatively, if the flux is zero or near-zero (i.e. below the threshold), we can say that there is no significant emission present within that area, and the mobile sensor can move to the next equipment pad for inspection (see the example in Fig. 3.5). This approach can also be extended toward quantifying an emission source [4]. We will see more in Chapter 4.

Control volume approach efficiently identifies whether an area contains an emission source before investing in fine-scale localization. There is no need to enter restricted areas; measurements can be taken from the perimeter. This strategy also works with different mobile platforms (e.g., drones, ground-based robots, handheld devices) and can be adapted for methane, VOCs, and other trace gases. However, control volume method may be less effective in highly variable wind conditions. Accuracy is limited by the ability to resolve small flux differences. Buildings and structures can distort wind patterns, making flux calculations less reliable.

3.2.5 Back Trajectory Analysis

Heuristic methods for gas source localization leverage simplified, experience-based approaches that do not require a complete model of gas dispersion but instead rely on observable cues such as concentration spikes, prominent wind directions, or historical

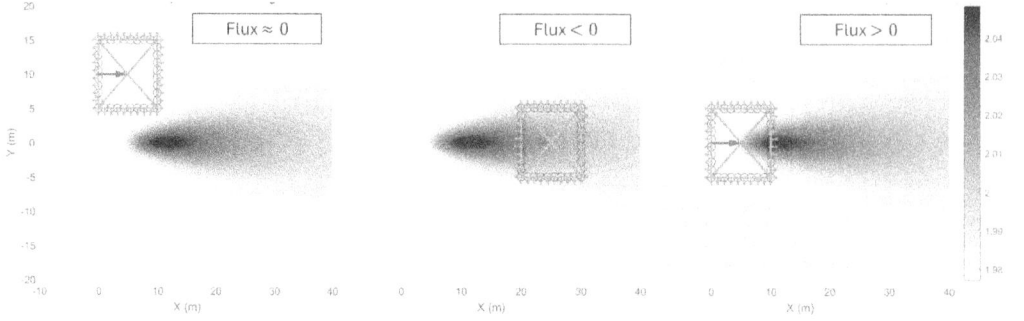

FIGURE 3.5 An example of the control volume-based localization method using the Gauss Divergence Theorem, a control volume path (box, showcasing the normal vectors) and measured wind (going from left to right). (left) A no source detected case, where the flux is approximately 0. (middle) A source detection with negative flux condition, where the source is not encapsulated inside the control volume and therefore is not localized. (right) A source detection with positive flux, indicating a successful source localization.

measurements in space. One such semi-heuristic approach involves the back trajectory analysis, where the sensor records concentration data at multiple locations, and wind information is used to estimate the source location through reverse propagation. By assuming that the trace gas in space, $\mathbf{x}(t)$, is released from an initial location subject to the wind speed and direction, $\mathbf{u}(x(t), t)$, in other words, advection, and is transported to a region detected by the sensor. This can be mathematically described by,

$$\frac{d\mathbf{x}(t)}{dt} = \mathbf{u}(x(t), t). \tag{3.13}$$

To simplify things, the wind vector measurement is often taken as a mean wind field approximation (i.e. the same wind speed and direction is applied across the spatial domain) and is assumed to be constant during some time interval, T, such that,

$$\hat{\bar{\mathbf{u}}} = \frac{1}{T} \int_{t_0}^{t_f} \mathbf{u}(x(t), t) dt \approx \bar{\mathbf{u}}. \tag{3.14}$$

If we solve the transport equation (3.13), we can see that the initial spatial location of the source and the initial time of release are unknown,

$$\mathbf{x}_f = \hat{\bar{\mathbf{u}}}(t_f - t_i) + \mathbf{x}_i. \tag{3.15}$$

The measurement location, $\mathbf{x}_m$, is in fact, the final position, $\mathbf{x}_f = \mathbf{x}_m$, and if the mean wind vector is reversed (i.e. projected 'backward') the source location can be assumed to be somewhere along the line. By taking a series of measurements at slightly different mean wind vectors, an optimization problem can be undertaken to triangulate the source,

$$\hat{\mathbf{x}}_i = \arg\min_{\mathbf{x}_i \in \Omega} \sum_{j=0}^{N} (\mathbf{x}_i - \mathbf{x}_{b_j, min})^2, \tag{3.16}$$

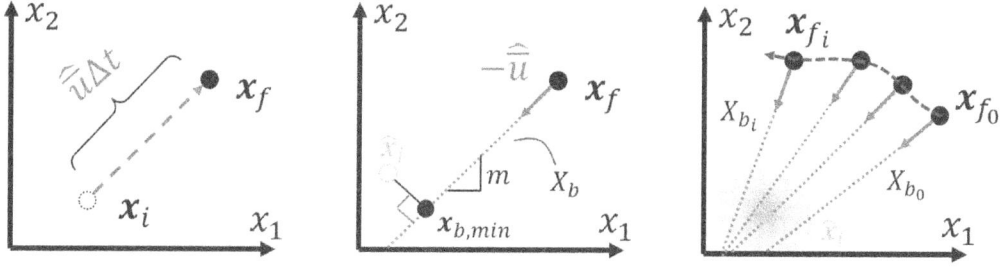

FIGURE 3.6 A diagram depicting the workflow for the back trajectory algorithm.

where $\mathbf{x}_{b_j,min}$ is the closest point along the j-th back trajectory line of discrete points $\mathbf{X}_b$. It turns out that this point is also the point where the line from the proposed $\mathbf{x}_i$ to $\mathbf{x}_{b_j,min}$ is perpendicular. The location $\mathbf{x}_{b,min}$ can be computed as

$$x_{b_{min,1}} = \frac{m(\hat{x}_{i,2} - B) + \hat{x}_{i,1}}{1 + m^2},$$

$$x_{b_{min,2}} = m x_{b_{min,1}} + B, \tag{3.17}$$

where $B = x_{f,2} - m x_{f,1}$ and m is the slope. An illustration is given in Fig. 3.6.

This method, often used in atmospheric studies, provides a way to infer source positions without requiring a real-time search strategy. Compared to direct source-seeking heuristics, back trajectory methods can be more robust to local concentration fluctuations but may require longer data collection periods and are sensitive to wind variability and measurement errors. In general, heuristic methods offer computational efficiency and adaptability, making them suitable for real-world deployment, but they typically lack the precision of model-based approaches and struggle in highly variable or low-information conditions.

Recent work by [23], developed a more robust version of this approach by defining something called an upwind survey region (USR) for assigning probable regions where the source could be located. The authors utilize a footprint function, $f_{x_m}(\cdot)$, to relate the concentration to the source rate,

$$C_{avg}(\mathbf{x}_m) = Q f_{\mathbf{x}_m}(\mathbf{x}_m). \tag{3.18}$$

They defined the USR as

$$USR(\mathbf{x}_m) = \{\mathbf{x} | Q_{min} f_{\mathbf{x}_m}(\mathbf{x}) > C_{min}\}. \tag{3.19}$$

If the average concentration measurement, $C_{avg}(x_m)$, is above a threshold, the region can be assigned as a region containing gas. Conversely, if the measurement is below the threshold, it is clear. However, because these systems are stochastic and there may be several measurements for a particular region, the authors proposed a probabilistic way to assign p-values to each of these regions. This approach is actually a combination of inverse dispersion modeling (by developing a library of footprint models), back trajectory analysis, and occupancy grid mapping.

TABLE 3.1 Summary of emission source localization strategies with their pros and limitations.

Strategy	Examples	Pros	Limitations
Direct Search	Infotaxis, Fluxotaxis, Entrotaxis, Spiral Surge casting, and anemotaxis	Real-time Adaptability, No Prior Knowledge Required, Fast Convergence in Favorable Conditions, Applicable in Unknown Environments	Susceptibility to Environmental Factors, Inefficiency in Weak Plumes, Local Minima Issues, Increased Search Time in Large Areas
Grid-based Search	Occupancy Grid Map, Gaussian Process Regression, Kernel DM+V, Kriging	Systematic and Structured Approach, Good for Mapping and Coverage, Less Affected by Environmental Variability, Useful for Persistent Emissions	Computationally Intensive, Inefficient for Fast Localization, Dependent on Sensor Density, Limited Adaptability
Inverse Dispersion and Inference	Lambert W STILT	Mathematically Rigorous, Utilizes Wind and Dispersion Models, Well-Suited for Sparse Data, Provides Uncertainty Estimates	Requires Accurate Models, Computational Complexity, Sensitive to Measurement Noise, Limited Adaptability to Dynamic Sources
Control Volume	Gauss Divergence Theorem	Fast Screening, Non-intrusive, Scalable, Adaptive to various gas types	Assumes Steady-State Conditions, Dependent on Sensor Sensitivity, Difficult in Urban/Complex Environments
Heuristic	Back Trajectory Analysis	Computationally Efficient, Adaptable	Assumes fixed source, Inefficient for Fast Changing Environment

3.3 OPEN CHALLENGES IN PRACTICE

Even though there are several localization methods available to us to apply, there are still scenarios and factors that are difficult to solve effectively for sUAS conducting localization tasks. Some example scenarios or factors include but are not limited to: multiple sources, required standoff distance, sUAS system disturbance, atmospheric stability, and sensor specific limitations.

To start, it is possible for multiple emission sources (or leaks) to be present around an oil and gas site, of which can contain multiple equipment pads with multiple components capable of leaking per pad. For example, many active sensor systems (e.g. TDLAS) only measure in situ concentration signals (i.e. in ppm), and as such, rely heavily on dispersion models or localization methods to identify emission sources. The act of localizing at a fine scale (sometimes referred to as 'pin pointing'), require exponentially increased flight times and investigation to squeeze the precision of the localization estimate. Furthermore, this estimate is in the form of GPS coordinates or a probabilistic distributions of points indicating where the likelihood of the leak source is. It does not however have the capability (...yet) to provide component-level localization/specification. Therefore, additional localization work is needed to determine the emission source attribution in the multi-source case.

Second, in many cases (e.g. oil and gas), sUAS operators are required to have a standoff distance from the equipment pads for intrinsic safety concerns. This can be thought of as an invisible bubble around the site or pad that prevents the sUAS operator from getting closer. This includes preventing the sUAS from flying directly over the equipment. While it is not clear whether the sUAS system is capable of providing a spark in open air from leaks that are typical of many sites, it is always better to 'err on the side of caution'. The result of this standoff distance is the decreased localization precision in grid and direct search methods.

Third, the sUAS system can often provide additional disturbances to the air surrounding the sUAS and conversely, the way the sensor receives the measurement signal. As mentioned in Chapter 2, the sensor placement is critical for improving the emission detection on-board the sUAS. A simple example would be the difference in disturbance by using a pumped-TDLAS system versus an in situ TDLAS (see Fig. 2.4). In the pumped system, the propeller disturbance is minimal at the point where the emission is being measured. Therefore, the localization in this context can be more accurate than with an in situ system flying low and slow to the ground.

Fourth, the atmospheric stability plays a key role in localizing an emission source. In the ideal case (with no obstructions), if the conditions are stable, moving in one direction, at one speed, with little turbulence or meandering, localization can become more of a trivial problem – easily solved with the localization methods described in this chapter. In practice, we often do not get the opportunity to measure emissions without obstructions or complex terrain. Additionally, the wind tends to always meander (i.e. change directions at varying rates), as well as speed up and slow down. During the localization task the conditions may vary in favorable and unfavorable ways. For example, in a complex environment and unstable atmospheric conditions, the wind field becomes very unpredictable and non-linear – making it extremely difficult to apply localization methods. sUAS operators may attempt to wait for the conditions to improve before conducting additional flights, but during the next flight the conditions may change, compromising the results. Alternatively, successive flights with different wind directions can be used to more accurately localize a source.

Pause and Reflect

How would prior knowledge of potential leak or emission sources and their locations help to inform the location estimate? What about scenarios that do not have any priors?

3.4 CHAPTER SUMMARY

In this chapter we learned about the source localization problem and the three general types of localization methods (i.e. direct search, grid-based, and inference-based). The direct search methods can be applied with little knowledge of the environment as they may utilize gradient information to find the source. We learned there are several methods to compute the gradient (e.g. from an information standpoint, entropy, or mass divergence, etc.). Grid-based methods, utilize various techniques to assign concentrations or probabilities to spatial grid points. These methods usually need the sUAS (in this case) to systematically sweep (back and forth) across the domain to develop the map. The result is a map that can be further processed to find the location. Inverse dispersion and inference methods rely on more model based and expensive CFD or Lagrangian approaches to simulate the dispersion of the emission source. Through optimization or other approaches, the localization can be carried out. An example was utilizing the Lambert W function to systematically estimate the location of the source through optimization. Additionally, we illustrated that these methods can be combined with search to create automated techniques for finding, localizing, and even quantifying the emission source. A control volume method and back trajectory analysis method were outlined, showcasing some simple, yet powerful ways to quickly identify emission sources and localize them. However, with each type of localization methods described here, there are general drawbacks related to the rapid changing of the weather conditions and/or emission source. Understanding when to use and when not to use these methods is key in developing rigorous protocols for localizing emissions.

Bibliography

[1] Richard G Baraniuk. Compressive sensing [lecture notes]. *IEEE Signal Processing Magazine*, 24(4):118–121, 2007.

[2] Horatio Scott Carslaw. J. c. Jaeger. *Conduction of Heat in Solids*, 2, 1959.

[3] Lance E Christensen. Miniature tunable laser spectrometer for detection of a trace gas, June 2017. US Patent 9,671,332.

[4] Stephen Conley, Ian Faloona, Shobhit Mehrotra, Maxime Suard, Donald H Lenschow, Colm Sweeney, Scott Herndon, Stefan Schwietzke, Gabrielle Pétron, Justin Pifer, et al. Application of Gauss's theorem to quantify localized surface emissions from airborne measurements of wind and trace gases. *Atmospheric Measurement Techniques*, 10(9):3345–3358, 2017.

[5] Claude R Dietrich and Garry N Newsam. Fast and exact simulation of stationary Gaussian processes through circulant embedding of the covariance matrix. *SIAM Journal on Scientific Computing*, 18(4):1088–1107, 1997.

[6] Jay A Farrell, John Murlis, Xuezhu Long, Wei Li, and Ring T Cardé. Filament-based atmospheric dispersion model to achieve short time-scale structure of odor plumes. *Environmental Fluid Mechanics*, 2(1-2):143–169, 2002.

[7] Tierney A Foster-Wittig, Eben D Thoma, and John D Albertson. Estimation of point source fugitive emission rates from a single sensor time series: A conditionally-sampled Gaussian plume reconstruction. *Atmospheric Environment*, 115:101–109, 2015.

[8] Tierney A Foster-Wittig, Eben D Thoma, Roger B Green, Gary R Hater, Nathan D Swan, and Jeffrey P Chanton. Development of a mobile tracer correlation method for assessment of air emissions from landfills and other area sources. *Atmospheric Environment*, 102:323–330, 2015.

[9] Derek Hollenbeck, Kevin Zheng, Demitrius Zulevic, and YangQuan Chen. Swarm robotic source seeking with fractional fluxotaxis. In *Proc. of the 2023 International Conference on Fractional Differentiation and Its Applications (ICFDA)*, 2023.

[10] Derek Hollenbeck, Demitrius Zulevic, and YangQuan Chen. A modified near-field Gaussian plume inversion method using multi-sUAS for emission quantification. In *2022 International Conference on Unmanned Aircraft Systems (ICUAS)*, pages 1620–1625. IEEE, 2022.

[11] Michael Hutchinson, Cunjia Liu, and Wen-Hua Chen. Information-based search for an atmospheric release using a mobile robot: Algorithm and experiments. *IEEE Transactions on Control Systems Technology*, 27(6):2388–2402, 2018.

[12] Michael Hutchinson, Hyondong Oh, and Wen-Hua Chen. Entrotaxis as a strategy for autonomous search and source reconstruction in turbulent conditions. *Information Fusion*, 42:179–189, 2018.

[13] Achim J Lilienthal, Matteo Reggente, Marco Trincavelli, Jose Luis Blanco, and Javier Gonzalez. A statistical approach to gas distribution modelling with mobile robots-the kernel DM+v algorithm. In *Proc. of the 2009 IEEE/RSJ International Conference on Intelligent Robots and Systems*, pages 570–576. IEEE, 2009.

[14] JC Lin, Christoph Gerbig, SC Wofsy, AE Andrews, BC Daube, KJ Davis, and CA Grainger. A near-field tool for simulating the upstream influence of atmospheric observations: The stochastic time-inverted Lagrangian transport (STILT) model. *Journal of Geophysical Research: Atmospheres*, 108(D16), 2003.

[15] Krithika Manohar, Bingni W Brunton, J Nathan Kutz, and Steven L Brunton. Data-driven sparse sensor placement for reconstruction: Demonstrating the benefits of exploiting known patterns. *IEEE Control Systems Magazine*, 38(3):63–86, 2018.

[16] Jörg Matthes, L Groll, and Hubert B Keller. Source localization by spatially distributed electronic noses for advection and diffusion. *IEEE Transactions on Signal Processing*, 53(5):1711–1719, 2005.

[17] Arye Nehorai, Boaz Porat, and Eytan Paldi. Detection and localization of vapor-emitting sources. *IEEE Transactions on Signal Processing*, 43(1):243–253, 1995.

[18] Surya G Nurzaman, Yoshio Matsumoto, Yutaka Nakamura, Kazumichi Shirai, Satoshi Koizumi, and Hiroshi Ishiguro. From Lévy to Brownian: a computational model based on biological fluctuation. *PLoS One*, 6(2):e16168, 2011.

[19] OFT Roberts. The theoretical scattering of smoke in a turbulent atmosphere. *Proceedings of the Royal Society of London. Series A, Containing Papers of a Mathematical and Physical Character*, 104(728):640–654, 1923.

[20] Eric Schulz, Maarten Speekenbrink, and Andreas Krause. A tutorial on Gaussian process regression: Modelling, exploring, and exploiting functions. *Journal of Mathematical Psychology*, 85:1–16, 2018.

[21] Dieter Fox Sebastian Thrun, Wolfram Burgard. *Probabilistic Robotics*. The MIT Press, 2006.

[22] Adil Shah, Grant Allen, Joseph R Pitt, Hugo Ricketts, Paul I Williams, Jonathan Helmore, Andrew Finlayson, Rod Robinson, Khristopher Kabbabe, Peter Hollingsworth, et al. A near-field Gaussian plume inversion flux quantification method, applied to unmanned aerial vehicle sampling. *Atmosphere*, 10(7):396, 2019.

[23] Witenberg SR Souza, Alexander J Hart, Benedito JB Fonseca, Mansour Tahernezhadi, and Lance E Christensen. A framework to survey a region for gas leaks using an unmanned aerial vehicle. *IEEE Access*, 12:1386–1407, 2023.

[24] E. Thoma and B. Squier. Draft Other Test Method 33A: Geospatial measurement of air pollution, remote emissions quantification - direct assessment (GMAP-REQ-DA). *Environmental Protection Agency (EPA)*, 2014. https://www3.epa.gov/ttnemc01/prelim/otm33a.pdf.

[25] Massimo Vergassola, Emmanuel Villermaux, and Boris I Shraiman. 'infotaxis' as a strategy for searching without gradients. *Nature*, 445(7126):406–409, 2007.

[26] Hans Wackernagel. Ordinary Kriging. In *Multivariate Geostatistics*, pages 79–88. Springer, 2003.

[27] Dimitri Zarzhitsky, Diana F Spears, William M Spears, and David R Thayer. A fluid dynamics approach to multi-robot chemical plume tracing. In *Proceedings of the Third International Joint Conference on Autonomous Agents and Multiagent Systems-Volume 3*, pages 1476–1477. IEEE Computer Society, 2004.

CHAPTER 4

Emission Source Quantification

Now that we have explored detection and localization in previous chapters, we will now turn our attention on quantification. In industry (e.g. oil and gas), this is referred to leak detection and quantification (LDAQ). In [57], the authors highlight advanced LDAQ methods in the context of sUAS – much of which is covered here. One of the first things to note is that there are several types of quantification strategies that we can choose from (simulation-, optimization-, mass-balance-, imaging-, and correlation-based). In some cases, the strategy involves prior knowledge about the location of the source, and in others, the source location is not generally needed or estimated in the process. In the following sections, we will highlight these different types of quantification and focus more on some key methods specifically for sUAS. Then, we will provide a high-level assessment of the different quantification methods in terms of complexity, precision, and cost.

4.1 TYPES OF QUANTIFICATION APPROACHES

In the literature we can find many different quantification approaches that depend on a range of assumptions for a chosen mode of sensing. The different types of quantification approaches can fall broadly into five general categories: simulation-, optimization-, mass-balance-, imaging-, and correlation-based. Although some of the methods introduced here are not directly applicable to sUAS-based quantification, they are helpful in understanding the quantification method landscape for future implementation or adaptations. The primary focus of this section will be to highlight these methods and give an introduction to sUAS methods. An interested reader can check out [57, 73, 84].

4.1.1 Simulation

A **simulation-based** quantification uses atmospheric dispersion models to estimate emissions by simulating the transport and diffusion of methane under given environmental conditions in forward or backward time integrations. For instance, if a

DOI: 10.1201/9781003669470-4

simulation is used to model or predict emission rate, or is used to predict the like-lihood of the rate, this can be largely referred to as a simulation-based approach. Simulation approaches may require a forward or backward time-stepping scheme to achieve the result but are mostly computed numerically.

Techniques like Backward Lagrangian Stochastic (bLS) and Stochastic Time-Inverted Lagrangian Transport (STILT) trace particle trajectories backward from sensor measurements to infer probable source locations and strengths. These methods rely on accurate meteorological data and turbulence models, making them powerful for regional-scale studies but sensitive to errors in wind field assumptions. This also includes simulations governed by partial differential equations (PDE) or even hybrid approaches that include a mixture of PDEs, stochastic equations, data-driven, and/or machine learning. Examples of the latter are with modern digital twins, which are sometimes very complex and implemented at a range of length-scales. We will see more in Chapter 8.

In application, for example, the accepted backward modeling approach used in the draft OTM-33A document [95] and in several applications (e.g., Dairy Farm [9], etc.) is the backwards Lagrangian stochastic (bLS) approach by [39]. The bLS approach aims to answer the general questions: *What is the proper form of the LS trajectory model?* As well as, *how can source estimates be extracted from the particle's backward LS trajectory?* This method provides a source estimation for an area source given the source location (with unknown source rate), assuming horizontally uniform surface source, and that the atmosphere is in horizontal equilibrium (see Fig. 4.1).

The Stochastic Time-Inverted Lagrangian Transport (STILT) model offers significant advantages over the Backward Lagrangian Stochastic (bLS) approach for source location estimation. Unlike bLS, which assumes steady-state turbulence and is best suited for homogeneous terrain, STILT incorporates variable meteorological fields, making it effective in complex environments [70]. By using an ensemble approach, STILT captures atmospheric transport uncertainty, improving robustness in

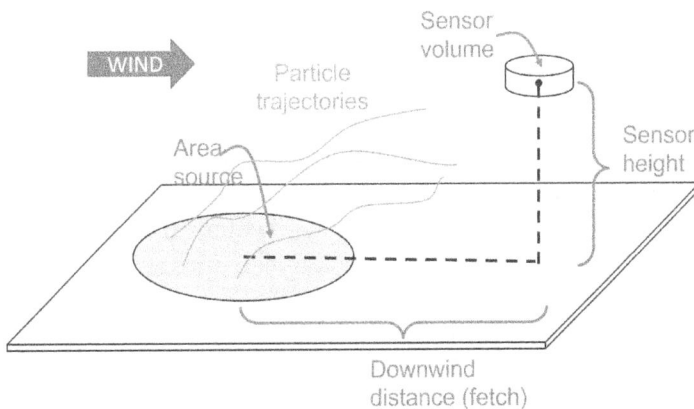

FIGURE 4.1   A diagram depicting the bLS approach (see [39]). ©**American Meteorological Society**. Used with permission.

source attribution. It also integrates mesoscale meteorological data (e.g., Weather Research and Forecasting (WRF), European Centre for Medium-Range Weather Forecasts (ECMWF)), reducing reliance on local wind measurements and enabling regional-scale source estimation [76]. Additionally, STILT's adaptive particle release mechanism enhances sensitivity to both near-surface and elevated sources, making it well-suited for multi-source emission studies [38]. These advantages make STILT a more flexible and accurate tool for methane emission quantification and localization. For example, in [64], researchers utilized the STILT model, combined with spatially distributed column measurements using FTIR instruments, to estimate surface fluxes with a Bayesian inversion-based framework.

Backwards Lagrangian Stochastic

A key simulation based approach for quantification of an emission source is the backwards Lagrangian Stochastic (bLS) approach. In brief, the method utilizes a forward model to describe the dynamics of the plume evolution, in the Lagrangian sense, and a backwards model to describe the dynamics of the observed measurements from the sensor to the source – in the backward time frame $(t' = -t)$. The forward model, formulated as a generalized Langevin equation, is evolved jointly as a Markov process,

$$du_i = a_i(\mathbf{x}, \mathbf{u}, t)dt + b_{i,j}(\mathbf{x}, \mathbf{u}, t)d\xi_j, \quad dx_i = u_i dt, \tag{4.1}$$

where the particle position is given by $\mathbf{x} = (x_1, x_2, x_3)$ and $d\xi_j$ is a random increment governed by a Gaussian process. The functions a_i and $b_{i,j}$ have to be specified such that the velocity probability density function, $g_a(\mathbf{x}, \mathbf{u}, t)$, satisfies the Fokker–Planck equation (FPE) [39],

$$\frac{\partial g_a}{\partial t} = \frac{\partial}{\partial x_i}(u_i, g_a) - \frac{\partial}{\partial u_i}[a_i(\mathbf{x}, \mathbf{u}, t)g_a] + \frac{\partial}{\partial x_i}[B_{i,j}(\mathbf{x}, \mathbf{u}, t)g_a]. \tag{4.2}$$

4.1.2 Optimization

An **optimization-based** quantification method applies inverse modeling techniques to minimize the difference between observed and predicted concentrations, identifying the most probable parameters and conversely estimating the emission rate and location. These are a class of simulation approaches that utilize a model to quantify emission based on optimization. These types of optimizations typically incorporate fast running closed-form models (e.g. the Gaussian plume model) to estimate the source terms (i.e. location, emission rate, dispersion parameters, etc.) and are classified in this book as optimization-based methods.

Gaussian Plume Model – A simple Digital Twin representation of an emission source can be given by the Gaussian Plume Model (GPM). The GPM can be mathematically described by

$$y = \mathcal{M}(\mathbf{x}; \theta) = \frac{Q}{\overline{U}}D_2 D_3, \tag{4.3}$$

where can be defined as $\theta = [Q, U, \mu_2, \mu_3, \sigma_2, \sigma_3]^T$. The dispersion functions are generally functions of the source location, downwind distance, and standard deviation of the plume. For example,

$$D_2(x_1, x_2, \mu_2, \sigma_2) = \frac{1}{\sqrt{2\pi}\sigma_2} \exp\left[-\frac{1}{2}\left(\frac{x_2 - \mu_2}{\sigma_2}\right)^2\right], \quad (4.4)$$

and

$$D_3(x_1, x_3, \mu_3, \sigma_3) = \frac{1}{\sqrt{2\pi}\sigma_3}\left(\exp\left[-\frac{1}{2}\left(\frac{x_3 + \mu_3}{\sigma_3}\right)^2\right] + \exp\left[-\frac{1}{2}\left(\frac{x_3 - \mu_3}{\sigma_3}\right)^2\right]\right), \quad (4.5)$$

where $\mu_3 = h + \Delta H$ is the effective height of the plume, h is the stack height, Q is the source rate in g/s, and ΔH is the plume rise [1]. The plume rise term can be calculated using models, e.g. Holland's formula which are dependent on temperature and wind speed. However, Holland's formula was designed for exhaust stacks and also takes into consideration the pressure, stack diameter, and temperature of the ambient air and stack. As a result, if the stack velocity is small, such as with sub-surface leaks that diffuse through the ground, the plume rise term tends toward zero. Therefore other factors should be considered, such as lapse-rate, which takes into consideration how the temperature changes as a function of altitude. Furthermore, for small leaks, even surface temperature may come into effect by introducing vertical momentum through natural convective forces. Holland found that there is a small correlation between the plume rise and the temperature gradient near the ground. Briggs mentioned that the plume rise is only dependent on the temperature gradient of the air to which the plume is rising. It is also mentioned that the gradient near the ground is not a good representation of the gradient at higher altitudes [19].

4.1.2.1 Point Source Gaussian – OTM33A

In [95], the point source Gaussian (PSG) is discussed in the context of the EPA's draft Other Test Method (OTM) 33A. The OTM33A method involves a vehicle equipped with a concentration measurement instrument (CMI, e.g. Picarro or LGR UGGA) that is parked downwind of the known source with the vehicle's engine turned off. While parked, the CMI collects concentration data at roughly 2.5 m above ground, at a known distance from the source, and simultaneously measures the variations in the wind direction using a sonic anemometer (e.g. a R.M. Young). Since the PSG calculation depends on enhanced concentration levels (i.e. the difference between the background concentration and the measured concentration), the background concentration signal can be calculated as the fifth percentile of the measured concentration timeseries signal. To calculate the PSG estimate, a simple 2-D Gaussian integration is used with no reflection term. The emission source rate estimate is then given as,

$$Q_E = 2\pi\sigma_2\sigma_3 U_m y_p, \quad (4.6)$$

where y_p is the peak concentration from the Gaussian fit, U_m is the mean wind speed, σ_3 and σ_2 are the vertical and lateral plume dispersion that can be determined from the meteorological conditions, such as the Pasquill–Gifford stability classification curves [59] (see Fig. 4.2). The accuracy of the OTM33A method is explored in [35, 36].

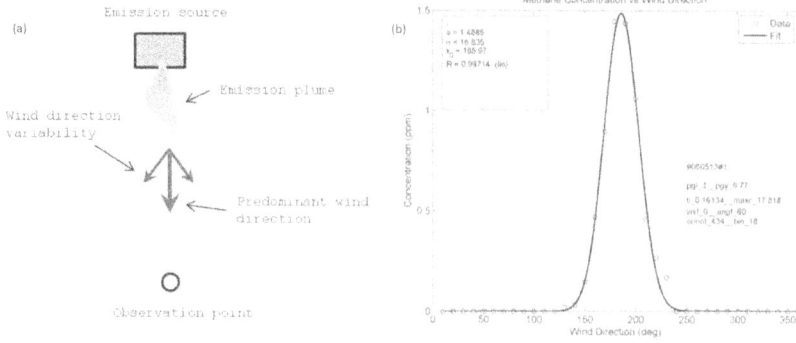

FIGURE 4.2 (a) Depiction of Gaussian plume dispersion with an observer making a stationary measurement downwind. (b) Resulting time-integrated data with a Gaussian fit applied [95].

4.1.2.2 Conditionally Sampled PSG (PSG-CS)

Building on the OTM33A method, [41] explores the use of conditionally sampled downwind measurements to estimate the emission source rate. To capture the ensemble mean of the downwind plume behavior, the GPM is used, where we denote the downwind distance, crosswind distance, and vertical position, respectfully, as $\mathbf{x} = [x_1, x_2, x_3]^T$. The crosswind plume center is given as $\mu_2 = 0$. The GPM is defined as a function of downwind distance and dispersion factors $D_2(x_1, x_2)$ and $D_3(x_1, x_3)$, given as,

$$y_m(\mathbf{x}) = \frac{Q}{U} D_2(x_1, x_2) D_3(x_1, x_3). \tag{4.7}$$

This method essentially aims to determine the emission source rate, Q, using the conditional mean concentration data, y_m, of the downwind plume. The lateral dispersion downwind of a continuous point source can be shown to have a Gaussian distribution such that it can be represented as

$$D_2(x_1, x_2) = \frac{1}{\sqrt{2\pi}\sigma_2}[-\frac{1}{2}(\frac{x_2{}^2}{\sigma_2})]. \tag{4.8}$$

However, the vertical dispersion (assuming vertical eddy diffusivity and wind speed that scales vertically to a power law) can be formulated as a parameterized stretched exponential (originally expressed in [98]),

$$D_3 = D_3(x_1, x_3) = \frac{A}{\bar{x}_3} \exp\left[-(\frac{Bx_3}{\bar{x}_3})^s\right]. \tag{4.9}$$

The parameters $\bar{x}_3$, s, A, and B are functions of the atmospheric stability and downwind distance, x_1. A, and B can be described using the usual Gamma function, $\Gamma(\cdot)$ as

$$A = s\Gamma(2/s)[\Gamma(1/s)]^2, \tag{4.10}$$

$$B = s\Gamma(2/s)\Gamma(1/s). \tag{4.11}$$

The conditional averaged concentration can be calculated using

$$\langle y|\theta\rangle = \frac{1}{n}\sum_{\theta_i \in \Theta}^{n} y(\theta_i), \tag{4.12}$$

where the set $\Theta(\theta) = \{\theta_i : |\theta - \theta_i| < \Delta\theta/2, \forall i = 1, 2, ..., n\}$ and $\Delta\theta = 2°$. The basic idea is to capture the plume geometry in the crosswind direction, which is further used to derive the least squares source estimate

$$Q = \Big[\sum_{i=1}^{N} \frac{D_2 D_3}{\overline{U}}\langle y|\hat{d}_i\rangle\Big] / \Big[\sum_{i=1}^{N}(\frac{D_2 D_3}{\overline{U}})^2\Big]. \tag{4.13}$$

As shown in [41], the lateral dispersion can be determined in two ways: classically, using atmospheric stability (for constants a_2 and p_2) [34],

$$\sigma_2 = a_2 x_{3_0} 1.9\Big(\frac{L_x}{x_{3_0}}\Big)^{p_2}, \tag{4.14}$$

and by reconstructing the lateral dispersion,

$$\sigma_{x_2} = \sqrt{\frac{1}{N}\sum_{i=1}^{N} \underline{\hat{d}}_i}, \tag{4.15}$$

where the N is the number of values in $\langle y|\underline{\hat{d}}\rangle$, and $\underline{\hat{d}}$ are $\hat{d}$ values that are greater than the minimum concentration (i.e., background) and $\pm 40°$ off the plume center θ_p. The distance $\hat{d}$ is calculated as

$$\hat{d}(\theta) = L_x \sin(\theta - \theta_p), \tag{4.16}$$

with $\theta_p = \arg\max_\theta\langle y|\theta\rangle$ (see Fig. 4.3).

4.1.2.3 Recursive Bayesian PSG (PSG-RB)

In the work from [3, 50], a moving sensor is considered within a recursive Bayesian framework. This is done by modeling the measured point source concentration as a GPM in (4.3). Since the measurement is taken at closer distances to the source, the lateral dispersion is taken as a random function such that

$$\int_{-\infty}^{\infty} D_2(x_1, x_2)dx_2 = 1. \tag{4.17}$$

This can be advantageous for instantaneous plumes. The integrated lateral concentration can be written as

$$y^{x_2}(x_1, x_3) = \frac{Q}{\overline{U}} D_3(x_1, x_3). \tag{4.18}$$

The choice of the vertical dispersion D_3 (originally expressed in [98]) is that of a parameterized stretched exponential function as shown in (4.9). The lateral dispersion

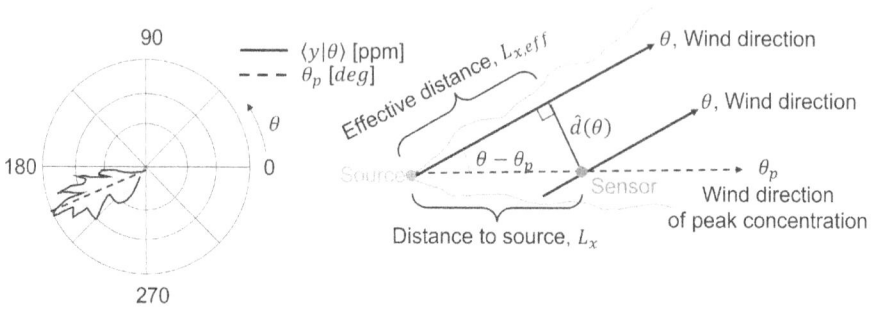

FIGURE 4.3 (left) Polar plot with the wind direction, θ as the radial axis, and the conditionally averaged concentration, $\langle y|\theta \rangle$ as the distance from the center. θ_p is the peak wind direction located at the maximum conditionally averaged concentration. (right) Illustration of the wind direction geometry for conversion of θ to crosswind position $\hat{d}$ with the source plume represented by the dashed lines. (reprinted from [41], with permission from Elsevier)

is given as (4.8). Then, by numerically integrating (4.7) and incorporating the vehicle movement V,

$$
y^{x_2} = \sum_{i=0}^{\infty} y(\mathbf{x}_i)\Delta t V = Q \sum_{i=0}^{\infty} \frac{\Delta t V}{\overline{U}_i} D_3(x_{1_i}, x_{3_i}) D_2(x_{1_i}, x_{2_i}). \tag{4.19}
$$

The recursive Bayesian approach described here is based on well pads and oil and gas production, which are used to help inform the path planning of the mobile sensor. For brevity, we will only cover the formulation of the quantification only. Starting with the definition of the posterior distribution,

$$
p(Q|M, W, \Lambda) = \frac{p(Q|W)p(M|Q, \Lambda)}{p(M|\Lambda)}, \tag{4.20}
$$

where M is the concentration data, W is the ancillary information (e.g. well pad characteristics), Λ is the meteorological conditions, $p(Q|W)$ is the prior, $p(M|Q, \Lambda)$ is the likelihood, and $p(M|\Lambda)$ is the evidence (which can be thought of as a normalization constant for the likelihood [105]). The prior is given as

$$
p(Q|W) = \frac{1}{\beta} \exp\left[-\left(1 + \gamma\frac{Q - \mu}{\beta}\right)^{-\frac{1}{\gamma}} \right]\left(1 + \gamma\frac{Q - \mu}{\beta}\right)^{-1-1/\gamma}, \tag{4.21}
$$

where the hyperparameters need to be fit to the application (for well-pad source, $\gamma = 1$, $\mu = 0.19$, $\beta = 0.23$ based on [16]). The likelihood function is chosen to be a Gaussian

$$
p(M|Q, \Lambda) = \frac{1}{\sqrt{2\pi}\sigma_e} \exp\left[-\frac{1}{2}\left(\frac{y^{x_2} - y^{x_2,M}}{\sigma_e}\right)^2 \right], \tag{4.22}
$$

where $y^{x_2,M}$ is the modeled concentration for a given source rate, and σ_e is the combined model and measurement error (outlined in [104]). The recursive approach involves replacing the prior with the previous posterior distribution found using the likelihood function,

$$p(Q|W)_i = \begin{cases} p(Q|W), & i = 1, \\ p(Q|M, W, \Lambda)_{i-1}, & i > 1. \end{cases} \qquad (4.23)$$

As the number of passes increases, the posterior distribution improves and can be used to estimate the source rate,

$$\hat{Q} = \int_{Q_{min}}^{Q_{max}} Q p(Q|M, W, \Lambda) dQ. \qquad (4.24)$$

Variations of this method were seen in [50], where the measurement noise was assumed to be Gaussian and also included a sUAS with sensor noise and utilized the flux plane mass balance method to estimate the source rate, which was used in the calculation of the posterior distribution. Further field tests of this method were carried out in [107].

4.1.2.4 Sequential Bayesian Markov Chain Monte Carlo PSG (PSG-SBM)

Utilizing the GPM for a likelihood function can also be applied in the context of a sequential Bayesian methods. In the work of [60], a sequential Bayesian Markov Chain Monte Carlo (MCMC) method was developed for a sUAS that scans the environment horizontally to update the posterior distribution of the estimated emission source terms. The emission source term parameters are denoted as Θ_k where $\Theta_k = [\mathbf{x}_s^T, Q_s, u_s, \phi_s, \zeta_s]^T$, the position is $\mathbf{x}_s$, source rate Q_s, wind speed and direction u_s and ϕ_s, and the model diffusion coefficients $\zeta_s = [\zeta_{s1}, \zeta_{s2}]^T$. The point source observations, $\mathbf{y}_{1:k} = \{y_1, y_2, ..., y_k\}$ are used within Bayes rule to update the posterior,

$$p(\Theta_{k+1}|\mathbf{y}_{1:k+1}) = \frac{p(\mathbf{y}_{k+1}|\Theta_{k+1}) p(\Theta_{k+1}|\mathbf{y}_{1:k})}{p(y_{k+1}|\mathbf{y}_{1:k+1})}. \qquad (4.25)$$

The likelihood model, $\mathcal{M}(\mathbf{x}_k, \zeta_k)$, in [60], based on observational data, $\overline{y}_k = \mathcal{M}(\mathbf{x}_k, \zeta_k) + \overline{v}_k$, was taken to be detection event, $p(\overline{y}_k|\Theta_k)$, if $y_k > y_{thr}$,

$$p(\overline{y}_k|\Theta_k) = \frac{1}{\sigma_k \sqrt{2\pi}} \exp\left[-\frac{(\overline{y}_k - \mathcal{M}(\mathbf{x}_k, \zeta_k))^2}{2\sigma_k^2}\right], \qquad (4.26)$$

and a non-detection event otherwise,

$$p(\underline{y}_k|\Theta_k) = \left(\frac{p_b}{2}\left[1 + \mathrm{erf}\left(\frac{y_{thr} - \mu_b}{\sigma_b\sqrt{2}}\right)\right]\right) + p_m + \left(\frac{p_s}{2}\left[1 + \mathrm{erf}\left(\frac{y_{thr} - (\mu_b + \mathcal{M}(\mathbf{x}_k, \zeta_k))}{\sigma_b\sqrt{2}}\right)\right]\right). \qquad (4.27)$$

The three terms in the non-detection event account for instrument noise, turbulence, and observing concentrations above the threshold, where $p_b + p_m + p_s = 1$, and μ_b and σ_b are mean background noise and standard deviation, respectively. Using a particle filter, the posterior can be approximated by a set of n weighted random samples $\{\Theta_k^{(i)}, w_k^{(i)}\}_{i=1}^n$,

$$p(\Theta_k|\overline{y}_{1:K}) \approx \sum_{i=1}^n w_k^{(i)} \delta(\Theta_k - \Theta_k^{(i)}), \qquad (4.28)$$

where δ is the Dirac delta function. The un-normalized weights are then updated using

$$\overline{w}_{k+1}^{(i)} = w_k^{(i)} \cdot p(y_{k+1}|\Theta_{k+1}^{(i)}). \tag{4.29}$$

Once the weights are determined, they can be normalized by dividing the summation of all the weights. Additionally, an effective sample size must be considered to avoid the degeneracy problem. The new samples undergo a MCMC step that is accepted with the likelihood probability distribution described earlier (see more details in [60]).

4.1.2.5 Near-field Gaussian Plume Inversion (NGI)

The near-field Gaussian plume inversion (NGI) method [88, 2] is a mass continuity model in principle, where the upwind and downwind concentration measurements, combined with wind measurements, of an emission source are subtracted to quantify emission flux. The NGI method is typically sampled around 100 m from the source. The sampling aims to capture the time-invariant behavior of the plume, which, under turbulent conditions, may not map out the characteristic Gaussian plume shape.

This is because it is assumed that spatial variability in the time-averaged plume is Gaussian. This method was initially carried out with a DJI S900 equipped with a ultra-portable greenhouse gas analyzer (UGGA) by Los Gatos Research Inc. (LGR). The flux estimate is derived by fitting the experimentally measured flux values, q_{me}, to the modeled flux values, q_{mo} given as,

$$q_{me} = (y - y_b)U(x_3)\rho, \tag{4.30}$$

where the modeled flux is given by the GPM in (4.3),

$$q_{mo} = \frac{Q_e}{2\pi\sigma_2(x_1)\sigma_3(x_1)} \exp\left(\frac{-(x_2 - \mu_2)^2}{2\sigma_2^2(x_1)}\right)\left(\exp\left(\frac{-(x_3 - \mu_3)^2}{2\sigma_3^2(x_1)}\right) + \exp\left(\frac{-(x_3 + \mu_3)^2}{2\sigma_3^2(x_1)}\right)\right). \tag{4.31}$$

The lateral and vertical dispersion relations are typically looked up in the Pasquill-Gifford (PG) stability tables, however, in this method, they are assumed to be linearly proportional to downwind distance,

$$\tau_2 = \sigma_2(x_1)/x_1, \quad \tau_3 = \sigma_3(x_1)/x_1. \tag{4.32}$$

Trying to solve (4.31) is not always well constrained, and thus the method proposes to separate (4.37) and fit the model along the x_3-direction,

$$q_{me,x_2} = q_{me} \frac{\tau_3 x_1 \sqrt{2\pi}}{\left(\exp\left(\frac{-(x_3 - \mu_3)^2}{2(\tau_3 x_1)^2}\right) + \exp\left(\frac{-(x_3 + \mu_3)^2}{2(\tau_3 x_1)^2}\right)\right)}. \tag{4.33}$$

The spatial variability in the x_3-direction has to be sampled to determine τ_3. The lateral spatial variability τ_2 and plume center μ_2 are determined simultaneously,

$$\mu_2 = \frac{\sum_j(q_{me,x_2} x_{2j})}{\sum_j(q_{me,x_2})}, \tag{4.34}$$

$$\tau_2 = \sqrt{\frac{\sum_j \left(q_{me,x_{2j}}\left(\frac{x_{2j}-\mu_2}{x_{1j}}\right)^2\right)}{\sum_j (q_{me,x_{2j}})}}. \tag{4.35}$$

Once the unknowns μ_3, τ_2, and μ_2 are determined, the source emission rate, $\hat{Q}$, can be estimated by minimizing the least squares fit between q_{me} and q_{mo}, parameterized by Q_e and τ_{x_3}. The uncertainty in $\hat{Q}$ and the impact of limiting τ_3 are given in [88].

4.1.2.6 Modified Near-field Gaussian Plume Inversion (mod-NGI)

To improve the optimization routine of the NGI method, the work in [58] introduced the modified NGI (mod-NGI) method. This modification looks at the conditioning of the parameter estimates by using the following flux-based likelihood function, represented by,

$$P(q|\theta) = D_2(x_1, x_2; \sigma_2, \mu_2)D_3(x_1, x_3; \sigma_3, \mu_3), \tag{4.36}$$

where $\theta = [Q, \mu_2, \mu_3, \sigma_2, \sigma_3]^T$ and the dispersion functions are outlined in Section 4.1.2.5. When the source rate is multiplied to (4.36), we have the modeled flux equation

$$q_{mo} = \mathcal{M}(\mathbf{x}, \theta) = QD_2(x_1, x_2; \sigma_2, \mu_2)D_3(x_1, x_3; \sigma_3, \mu_3). \tag{4.37}$$

Consider that experimental measurements do not give access to the above likelihood function but can rather give an estimate of it and let's assume an initial source rate estimate $\hat{Q}$ such that

$$\hat{P}(q_{me}|\hat{\theta}) \approx \frac{Q}{\hat{Q}} D_{x_2}(x_1, x_2; \hat{\sigma}_2, \hat{\mu}_2)D_3(x_1, x_3; \hat{\sigma}_3, \hat{\mu}_3). \tag{4.38}$$

The likelihood of observing the entire dataset then becomes,

$$\hat{P}(D|\hat{\theta}) = \Pi_{i=1}^{N}\hat{P}(q_{me}^i|\hat{\theta}). \tag{4.39}$$

Given that analytically solving for an optimizer using (4.5) can be complex, we make the assumption that the plume is far from the ground is only for the parameter initialization steps. This allows for the approximation of (4.5) by

$$D_{x_3}(x_1, x_3; \hat{\sigma}_3, \hat{\mu}_3) \approx \frac{1}{\sqrt{2\pi}\sigma_3(x_1)} \exp\left(\frac{-(x_3 - \mu_3)^2}{2\sigma_3^2(x_1)}\right). \tag{4.40}$$

Solving for the maximum log likelihood estimate (MLE) yields,

$$\hat{\tau}_2 = \sqrt{\frac{1}{Nx_1^2}\sum_{i=1}^{N}(x_{2i} - \mu_2)^2}, \quad \hat{\tau}_3 = \sqrt{\frac{1}{Nx_1^2}\sum_{i=1}^{N}(x_{3i} - \mu_3)^2}. \tag{4.41}$$

It can be readily apparent that the MLE is the definition of the standard deviation conditioned only on the spatial coordinates. To make this MLE conditioned on the flux measurements, one can substitute for the weighted standard deviation. For example, the horizontal scale factor is given by

$$\hat{\tau}_2 = \sqrt{\frac{N}{(N-1)x_1^2}\frac{\sum_{i=1}^{N} q_{me}^i(x_{2i} - \mu_2)^2}{\sum_{i=1}^{N} q_{me}^i}}. \tag{4.42}$$

In order to estimate τ_2 and τ_3, the plume centers need to be computed. This is done using the measured flux as weights and computing the center of mass,

$$\hat{\mu}_2 = \frac{\sum_{i=1}^{N} q_{me}^i x_{2i}}{\sum_{i=1}^{N} q_{me}^i}, \quad \hat{\mu}_3 = \frac{\sum_{i=1}^{N} q_{me}^i x_{3i}}{\sum_{i=1}^{N} q_{me}^i}. \tag{4.43}$$

Alternatively, the plume widths can be estimated directly without knowledge of plume location

$$\hat{\sigma}_2 = \sqrt{\frac{N}{(N-1)} \frac{\sum_{i=1}^{N} q_{me}^i (x_{2i} - \mu_2)^2}{\sum_{i=1}^{N} q_{me}^i}}, \tag{4.44}$$

and similarly for $\hat{\sigma}_3$. Once the dispersion and plume center parameters are estimated and an initial estimate of the source rate is established for $\hat{Q}$, the optimization of parameters θ can be undertaken, such that,

$$\hat{\theta} = \min_{\theta} \; J(\mathbf{X}, \theta), \quad J(\mathbf{X}, \theta) = \frac{1}{N} \sum_{i=1}^{N} (\mathcal{M}(\mathbf{x}_i, \theta) - q_{me}^i)^2, \tag{4.45}$$

where $\mathbf{X}$ is the observation data set trajectory defined as $\mathbf{X} = [\mathbf{x}_1, \mathbf{x}_1, \cdots, \mathbf{x}_N]$. This optimization can be carried out using the MATLAB *fminsearchbnd* function [32].

4.1.3 Mass-balance

A **mass-balance-based** quantification method aims to quantify emissions by measuring methane fluxes across control volumes or integrating concentrations and wind speeds over a defined cross-section. These methods rely on foundational equations for mass conservation and continuity (including the Gauss Divergence Theorem).

The mass balance approach aims to estimate an emission source by balancing the mass flux leaving or entering a control volume. Generally, there are two path planning approaches to the mass balance method: (1) rectangular vertical flux plane (or curtain) downwind of the source and (2) a cylindrical flux plane enclosing the source. For a well behaved plume under stable atmospheric conditions, the downwind plume contains all the flux. The sampling distance from the source may vary based on each submethod. The measured flux plane data can be sparse and is typically subject to spatial interpolation.

4.1.3.1 *Vertical Flux Plane (VFP)*

The flux plane method generally involves sampling within a plane, vertically or horizontally, upwind and downwind, of an emission source. It has been applied in several works [7, 6, 22, 21, 43, 54, 75, 81, 83, 89, 90, 102, 101, 100, 15]. The plane is typically sampled using a raster-scanning approach, capturing the plume within the width and height of the plane. The emission rate (in moles s^{-1}) can be estimated as

$$Q_c = \iint_{\Omega_p} n_{ij}(y - y_b)\mathbf{u} \cdot \hat{\mathbf{n}}_f dx_2 dx_3, \tag{4.46}$$

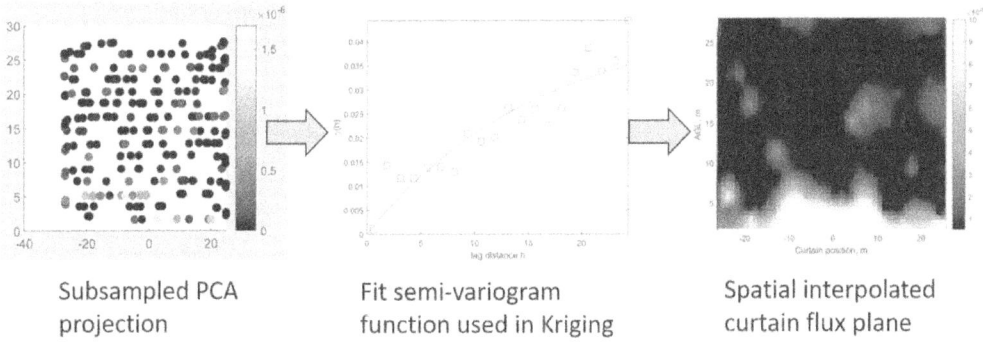

Subsampled PCA projection   Fit semi-variogram function used in Kriging   Spatial interpolated curtain flux plane

FIGURE 4.4 Demonstration of using sampled flux plane data and applying kriging to it for spatial interpolation [56].

where n_{ij} is the mole density of air (given standard temperature and pressure), $(y-y_b)$ is the enhanced mole fraction (referenced to air), y_b is the background mole fraction, $\mathbf{u}$ is the wind speed vector, and $\hat{\mathbf{n}}_f$ is the flux plane normal vector (see Fig. 4.4). Since the measurements are sparse, the integral is irregularly spaced. To combat this, the sparsely sampled points are spatially interpolated using techniques, such as inverse distance weighting (IDW) [28] or kriging [99]. This is a common problem in geostatistics to interpret unknown data, $y(\mathbf{x}_0)$, from desired spatial locations $\mathbf{x}_0$ in domain $\Omega \in \mathbb{R}^2$ (in our case is the domain of the plane $\Omega_p \in \mathbb{R}^2$), only using N sparse sampling points, $y(\mathbf{x}_i)$, based on some optimal weights, λ_i,

$$\hat{y}(\mathbf{x}_0) = \sum_{i=1}^{N} \lambda_i y(\mathbf{x}_i). \tag{4.47}$$

4.1.3.2 Cylindrical Flux Plane (CFP)

A variation to the VFP is the Cylindrical Flux Plane (CFP). This method has been used with manned aircraft as it is not as easy to raster-scan a rectangular flux plane. The methodology is essentially very similar to the VFP and can be found in the work by [85], omitted here for brevity.

4.1.3.3 Path-Integrated Vertical Flux Plane (PI-VFP)

A variation of the VFP is the path-integrated vertical flux plane (PI-VFP). This method utilizes a bs-TDLAS approach in that the instrument points straight down and scans or circles the emission source (see Fig. 4.5). In [44], the AVIRIS-NG manned aircraft used IMAP-DOAS technique to retrieve methane concentrations and estimated fluxes using a PI-VFP type calculation. This approach was compared with the GDT and Gaussian inverse approaches during a joint-flight campaign.

The emission rates were estimated by, $Q \approx \mathbf{u} \cdot \mathbf{n} \sum_i V_i \Delta s_i$, were V_i represents the vertically integrated concentration, and Δs_i is a path segment along the boundary. The individual measurements are integrated together (referred to as integrated mass enhancement (IME)) such that $\text{IME} = k \sum X_{CH_4}(i) \cdot S(i)$. The value X_{CH_4} is

Internal Leak (flux>0)   External Leak (flux=0)

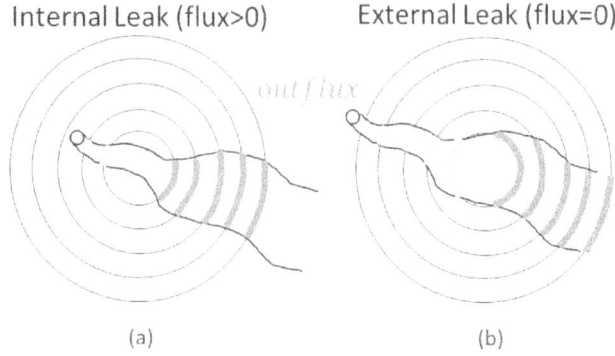

(a)   (b)

FIGURE 4.5 Example of the VFP-PI strategy via a sUAS sensing in circular trajectories with (a) being an internal leak producing a net positive flux and (b) being an external leak producing a net zero flux (see [103] for more details).

the methane plumes that exceed the minimum threshold of 200 ppm/m and k is a conversion factor.

Using a remote methane leak detector (RMLD) sensor fitted to a small quadrotor sUAS, a circular scanning approach can be applied to sample horizontally a site of interest. The sensor uses a bs-TDLAS to measure integrated methane emissions from a known height. The resulting measurements are then combined with wind measurements to estimate the flux [45, 103],

$$Q = \oint_{\mathcal{S}} \mathbf{u} \cdot \mathbf{n}_s (y - y_b) ds, \tag{4.48}$$

where $\mathbf{n}_s$ is the normal with respect to the path of travel such that $s \in \mathcal{S}$, y is the column measured concentration and y_b is the column background concentration. This calculation encompasses a single circular loop and if the source is encapsulated, multiple passes can be used to estimate the source. For instance, given n passes,

$$\hat{Q} = \frac{1}{n} \sum Q_i. \tag{4.49}$$

In practice, the circular flight path is actually made up of differential line segments that used to discretely solve (4.48). The emission source location can also be identified through course horizontal raster scanning over the area of interest to identify areas with high likelihoods. Once an area has been identified, additional flights can be conducted at finer spatial scales (using horizontal or pattern-free approach) to improve the resolution of the likely areas. The observations are then combined with triangular natural neighbor interpolation to determine the maximum observed concentration for the estimate of the emission source location.

4.1.3.4 Gauss Divergence Theorem

In the paper by [27], Conley et al. they focused on the continuity equation,

$$Q_c = \left\langle \frac{\partial m}{\partial t} \right\rangle + \iiint \nabla \cdot y\mathbf{u} dV, \tag{4.50}$$

where m is the mass of the aerosol, $\langle \cdot \rangle$ is the expectation or average, $y = Y + y'$ is the concentration (comprised of an average term and a deviation term), $\mathbf{u}$ is the wind speed, and V is the volume of the area of interest. The flux divergence can be expanded as,

$$\nabla \cdot y\mathbf{u} = \mathbf{u} \cdot \nabla y + y\nabla \cdot \mathbf{u}. \tag{4.51}$$

The surface integral is taken to be a cylinder, which can be broken into several parts: the floor, the walls of the cylinder, and the top. The height of the cylinder is taken such that the emission is encapsulated with in the minimum and maximum height. The resulting emission rate can be calculated as

$$Q_c = \left\langle \frac{\partial m}{\partial t} \right\rangle + \int_0^{z_{max}} \oint y'\mathbf{u}_h \cdot \hat{\mathbf{n}} dl dx_3, \tag{4.52}$$

where x_3 represents the altitude, and l the flight path. The temporal trend of the total mass $\left(\frac{\partial m}{\partial t}\right)$ within the volume can be estimated from the measurements. The cylinder passes can be vertically binned and discretely summed up,

$$Q_c = \frac{\Delta m}{\Delta t} + \sum_{x_3=0}^{x_3=x_{3,max}} \left(\sum_0^L (\rho \cdot u_n) \cdot \Delta s \right) \cdot \Delta x_3, \tag{4.53}$$

and ρ represents the scalar air density.

4.1.3.5 Vertical Flux Planes with GLM (GLM-VFP)

In [48], a 3D grid of airborne measurements are collected across multiple landfill sites. The resulting downwind observational points are then spatially interpolated with IDW and used to calculate the total mass flux (MF). The multiple steady-state Gaussian dispersion models,

$$y(\mathbf{x}) = \frac{Q}{2\pi\sigma_2\sigma_3 U} \exp\left(\frac{-x_2^2}{2\sigma_2^2}\right) \exp\left(\frac{-(x_3 - \mu_3)^2}{2\sigma_3^2}\right), \tag{4.54}$$

are applied to a fixed grid (50 m by 50 m), where the mixing ratios found over each individual landfill was used to calculate the model mass flux (MMF), for each site, integrated along the x_1, x_2, and x_3 directions). The experimental measurements are then used with simulation measurements and a general linear model,

$$\min_\alpha \left| MF - \sum_{i=1}^{max} (MMF_i \cdot \alpha_i) \right|, \tag{4.55}$$

to approximate the emission coefficient, α_i, from multiple landfill sources. The emission findings are further corroborated with a local eddy covariance tower measurement.

4.1.3.6 Micrometeorological Mass Difference (MMD)

Utilizing the technique from [31], sampling the plume far enough downwind of the source, the averaged MMD can be calculated as,

$$Q_{\overline{Uy}} = \overline{\iint U_{(x_2,x_3)}(y_{(x_2,x_3)} - y_b)dx_2dx_3} = \int \chi(x_3)dx_3, \tag{4.56}$$

where $U_{(x_2,x_3)}$ is the normal wind speed relative to the plane. The work in [47] utilized the time-average of the line-integral of the instantaneous product of U and y in the x_2-direction. Alternatively, while using a laser fetch, an instantaneous product of a single wind measurement U and line-averaged laser concentration was used,

$$\chi(x_3) \approx \Delta x_2 \overline{U_{(x_3)}(y_{(x_3)} - y_b)}. \tag{4.57}$$

This method can also be used to calculate the turbulent fluxes,

$$\frac{Q_{tur}}{Q} = \frac{(Q_{\overline{U}\,\overline{y}} - Q_{\overline{Uy}})}{Q_{\overline{Uy}}}, \tag{4.58}$$

where $Q_{\overline{Uy}}$ is calculated from the flux term in (4.57) and $Q_{\overline{U}\,\overline{y}}$ in (4.60),

$$Q_{\overline{U}\,\overline{y}} = \int \overline{U_{(x_3)}} \overline{\int (y_{(x_2,x_3)} - y_b)dx_2dx_3} = \int \chi(x_3)dx_3, \tag{4.59}$$

where

$$\chi(x_3) \approx \Delta x_2 \overline{U_{(x_3)}}\ \overline{(y_{(x_3)} - y_b)}. \tag{4.60}$$

This prescription of the flux does not capture the turbulent component of the horizontal flux (albeit wrong) is often necessary due to the short time-scale behavior of the wind (e.g., limitations in wind measurement devices).

4.1.3.7 *Vertical Radial Plume Mapping (VRPM)*

The vertical radial plume mapping approach (compared with other methods in [11]), utilizes a long path TDLAS instrument from the ground. The laser is aimed at retro-reflectors, situated perpendicular and downwind of the source. The height of the retro-reflector constitutes the different radial angles where the path-integrated concentrations are combined with the normal wind component to estimate the flux (similar to VFP or MMD). An illustration of this is seen in Fig. 1.2 (see [11] for more details).

4.1.4 Imaging

An **imaging-based** quantification method utilizes optical or spectroscopic techniques to visualize methane plumes and estimate emission rates based on plume characteristics and atmospheric conditions. The image-based methods rely on pixel-like column data, from hyperspectral or even from optical gas imaging, to quantify the emission. An example would be integrated mass enhancement method [63, 52].

In this section, we overview the imaging-based methodology for quantifying methane emissions. This typically includes techniques that sample images passively, such as TIR, MWIR, or other OGI-based instrumentation. The methods mentioned here that can quantify methane emissions are considered as quantitative optical gas imaging (QOGI).

4.1.4.1 Mid-Wave Infrared (MWIR) and Hyperspectral

As mentioned in Chapter 1, methane has a unique transmittance or absorption spectrum. The goal of mid-wave infrared (MWIR) cameras and hyperspectral cameras are to extract the methane signal from the absorption characteristics. In the work by [86], the detection limits of MWIR band of a hyperspectral data was explored using the Spatially-Enhanced Broadband Array Spectrograph System (SEBASS) airborne instrument. They also provided a comparison between LWIR and MWIR (see [86] for more details) using the radiative transfer model

$$R_s = (R_T^\uparrow + R_S^\uparrow) + t\{\epsilon_s B(T_s) + (1 - \epsilon_s)[\frac{R_T^\downarrow + R_S^\downarrow}{1 - S(1 - \epsilon_s)}]\}, \qquad (4.61)$$

where R_s is the total radiance at the sensor, $R_T^\uparrow$ is the upwelling emitted atmospheric path radiance, $R_T^\downarrow$ is the downwelling emitted atmospheric path radiance, $R_S^\uparrow$ is the scatter path radiance at the sensor, $R_S^\downarrow$ total solar radiance that reaches the surface, t is the atmospheric transmittance, ϵ_s is the surface emissivity, and $B(T_s)$ is the blackbody radiation at the surface temperature.

Other works, such as [17], have used MWIR cameras combined with two Pergam Methane Mini G lasers in pipeline leak detection. In [33], a FLIR GF320 and a RMLD were used together to make volumetric flow rate calculations in the laboratory using a data fusion approach. In [94], they utilized a thermal camera and steady-state energy balance approach to estimate methane emissions from thermal anomalies in urban landfills.

The last method in this category we will discuss is the iterative maximum a posteriori differential optical absorption spectroscopy method (IMAP-DOAS). The IMAP-DOAS method was applied to the AVIRIS-NG [96, 97] aircraft and measures reflected solar radiation between 0.35 μm and 2.5 μm with 5 nm spectral resolution and sampling. Using a nonlinear iterative minimization of the differences between modeled and measured radiance. The measured concentrations can be applied to the PI-VFP method to calculate fluxes [44]. Variations in this approach for retrieving methane concentrations has been seen in [40] for albedo correction and [69] anomaly-based mass balance. For example, in [40], a fast iterative shrinking thresholding algorithm (FISTA) was applied instead of the IMAP-DOAS algorithm to improve the speed and accuracy of retrieval. The authors accomplished this by linearizing Beer-Lambert Law,

$$L = L_0 \exp(-s\alpha) \approx L_0 - \alpha s L_0, \qquad (4.62)$$

to estimate the concentration map of methane, α, using the absorption bands, s, from that of the background signal, and L_0 in the ℓ_1 regularization sense (see Chapter 8).

4.1.5 Correlation

A **correlation-based** quantification method aims to determine emission rates by analyzing statistical relationships between observed methane concentrations and environmental variables (such as eddy covariance) or with tracer gas observations. There are a few methods that utilize correlation in quantifying emissions. Here we discuss two primarily.

4.1.5.1 Tracer Correlation (TCM)

The tracer correlation method, or isolated source tracer ratio method, initially proposed and implemented in works by Lamb et al [66] and Czepiel et al [29], aims to quantify the emission rate of an unknown gas species by releasing a tracer gas at a known flow rate while measuring both the tracer and the unknown signals collocated downwind. This method assumes that the location of the source is known and, at the measurement location, the plume is well mixed. The elevated signal downwind also needs to typically be greater than 50 ppb. The authors report uncertainty estimates of ±15%. The general equation is given as

$$Q_m = Q_t \frac{y_m}{y_t}, \tag{4.63}$$

where Q_t is the tracer release rate, and y_m and y_t are the elevated mixing ratios of the unknown source gas and tracer gas, respectively. A comparison study between TCM and other fugitive emission quantification methods are studied in [11]. The effect of wind on accuracy of the TCM was explored for landfills using WRF model [62]. An in situ method was used to evaluate the collection efficiency of gas extraction wells based on tracer gas [61].

Variations of the quantification of TCM were explored in [74], which quantified emission rates based on the plume integration of a transect, peak height of the transect using a scatter plot to calculate the ratio (best fit line), and comparison with fitted Gaussian plume model. A landfill field comparison of methane emission models were compared to measured emissions using TCM [30]. The TCM method was also applied to quantifying emissions from dairy farms in [9].

A dual tracer method was explored in [84]. The second tracer provides for closer downwind measurements that can be refined by assessment of plume position, as well as in the far-field measurements the second tracer becomes an internal standard to the measurement. A mobile version of the TCM approach was proposed in [42].

4.1.5.2 Eddy Covariance (EC)

The eddy covariance method aims to estimate the emission flux from a footprint area given the boundary layer meteorology. Historical developments and current implementations of this method are summarized in [53]. This method generally assumes stationarity of the measured data and fully developed turbulent conditions [49]. One way it can be expressed as

$$Q = \frac{1}{t_f - t_i} \int_{t_i}^{t_f} (y(t) - \overline{y})(w(t) - \overline{w})dt, \tag{4.64}$$

where the time-averaged concentration and vertical wind speed is $\overline{y}$ and $\overline{w}$, respectively. There are several assumptions required to make this flux calculation (for more details see [20]).

4.2 IMPLEMENTING sUAS-BASED QUANTIFICATION

Depending on the types of quantification methodologies used, several different implementation details can be leveraged to carry out the task, such as, spatial interpolation of the data, or controlling multiple sUAS in a coordinated manne. Here we highlight several techniques used in mass-balance methods and multi-sUAS deployment.

4.2.1 Interpolation and Data Driven Techniques

4.2.1.1 Kriging

In ordinary kriging [99], a semivariogram is used to model the spatial variability and, given a spatial distance, h, is defined as,

$$\hat{\gamma}(h) = \frac{1}{2N(h)} \sum_{i=1}^{N(h)} (y(\mathbf{x}_i) - y(\mathbf{x}_i + h))^2. \tag{4.65}$$

This experimental semivariogram can be fitted to the model semivariogram with one of several common functions: circular, spherical, exponential, Gaussian, or linear. The weights are determined by solving

$$\sum_{j=1}^{N} \lambda_j C(\mathbf{x}_i - \mathbf{x}_j) + \mu(\mathbf{x}_0) = C(\mathbf{x}_i - \mathbf{x}_0), \text{ for } i = 1, 2, ..N, \tag{4.66}$$

where $C(\cdot)$, in this context, represents the point support covariance matrix. This matrix is related to the semivariogram, $\gamma(h) = C(0) - C(h)$ [99], and the mean squared prediction error is $\sigma_e^2 = Var(y(\mathbf{x}_0) - \hat{y}(\mathbf{x}_0))$, which, for ordinary kriging, is minimized to make the estimated values $\hat{y}(\mathbf{x}_0)$ optimal. Furthermore, the estimator should be unbiased (e.g. $E[\hat{y}(\mathbf{x}_0)] = E[y(\mathbf{x}_0)]$), which requires $\sum \lambda_i = 1$ and the spatial mean to be stationary $E[y(\mathbf{x})] = \mu$, $\forall \mathbf{x} \in \Omega$.

If the kriging process is not stationary, it is considered, at best, an approximate solution to the spatial interpolation problem and incorrect at worst. A better approach could be to apply a spectral method that takes into consideration non-stationarity and higher frequencies, namely, the high frequency kriging method [46]. Consideration of temporal observations could be included as well, see quantile kriging in [68].

Available tools, such as Kriging Assistant (KA) [72], Golden Software Surfer, or ESRI Geostatistical Analyst for ArcMap have been used in the literature before. For irregular geographical units with different sizes and shapes, the interested reader should consult [51].

4.2.1.2 Inverse Distance Weighting

Given a set of known data points $\{\mathbf{X}, \mathbf{Y}\}$, where the measured values are denoted as $\mathbf{Y} = [y_1, y_2, \ldots, y_N]^T$ and measurement locations as $\mathbf{X} = [\mathbf{x}_1, \mathbf{x}_2, \ldots, \mathbf{x}_N]^T$. The term y_i represents the measured value at location $\mathbf{x}_i = [x_{i,1}, x_{i,2}]$, the estimated value $\hat{y}(\mathbf{x})$

at an unknown point $\mathbf{x}$ is computed as

$$\hat{y}(\mathbf{x}) = \frac{\sum_{i=1}^{N} w_i y_i}{\sum_{i=1}^{N} w_i}, \tag{4.67}$$

where w_i is the weight assigned to each known data point and is given by,

$$w_i = \frac{1}{d_i^p}, \tag{4.68}$$

where d_i is the Euclidean distance between the unknown point $\mathbf{x}$ and the known point $\mathbf{x}_i$, given by

$$d_i = \|\mathbf{x} - \mathbf{x}_i\| = \sqrt{(x_1 - x_{i,1})^2 + (x_2 - x_{i,2})^2}. \tag{4.69}$$

p is the power parameter that controls the influence of distance. A higher p value gives more weight to closer points, making the interpolation more localized. A common choice is $p = 2$, which is known as inverse squared distance weighting [92]. An enhanced version of the IDW was proposed in [71] to include an adaptive distance-decay parameter based on the density characteristics of the sampled points.

Unlike kriging or other geostatistical methods, IDW does not require variogram modeling or statistical assumptions. IDW ensures that known data points are preserved, meaning the estimated value at a measurement location is exactly the measured value. Closer points contribute more to the interpolated value, while farther points contribute less. The power parameter p and the number of nearest neighbors N used in interpolation can be adjusted to optimize accuracy.

IDW is generally simple and computationally efficient. It is easy to implement and requires less computational power compared to kriging or machine learning approaches. It produces smooth spatial fields, making it useful for applications where abrupt changes are not expected. IDW does not assume stationarity or a specific distribution of the data.

However, IDW has its limitations. For example, the selection of the power parameter p is empirical and requires tuning for different datasets. Depending on how many points are included in the interpolation, IDW may not capture localized trends well. IDW may not perform well at the boundaries of a dataset since there are fewer data points influencing estimates near the edges. A single extreme value in the dataset can significantly affect interpolated results.

4.2.1.3 Kernel DM+V

The gas distribution modeling problem aims to address the issue of learning a predictive model measurement at a specified query point given past measurements in space

$$p(y|\mathbf{x}, \mathbf{x}_{1:n}, y_{1:n}). \tag{4.70}$$

The sensor is modeled as $r \in [0, 1]$, with,

$$r_i = \frac{y_i - y_{min}}{y_{max} - y_{min}}, \tag{4.71}$$

such that $y_{min} = \min\{y_i\}_1^n$, and $y_{max} = \max\{y_i\}_1^n$. Then, using the $k = 1, 2, \ldots, N$ grid points in some defined domain $\Omega \in \mathbb{R}^2$. The weights can be computed using a kernel distribution with a kernel width. Here the kernel is represented with the Normal distribution

$$w^{(k)} = \sum_{i=1}^{N} \mathcal{N}(|\mathbf{x}_i - \mathbf{x}|, \sigma). \tag{4.72}$$

The weighted measurements are then computed as

$$R^{(k)} = \sum_{i=1}^{N} \mathcal{N}(|\mathbf{x}_i - \mathbf{x}|, \sigma) r_i. \tag{4.73}$$

A confidence map is then defined as

$$\alpha^{(k)} = 1 - \exp\left(-\left(\frac{w^{(k)}}{\sigma_\Omega}\right)^2\right) \tag{4.74}$$

to compute the updated grid cell measurements

$$r^{(k)} = \alpha^{(k)} \frac{R^{(k)}}{w^{(k)}} + (1 - \alpha^{(k)}) r_0, \tag{4.75}$$

where σ_Ω is a scaling term (which can be related to σ by setting $\sigma_\Omega = \mathcal{N}(0, \sigma)$, and r_0 is the average of points with low α (which can be computed as, $r_0 = \frac{1}{n} \sum_{i=1}^{n} y_i$). The variance map can be computed by first computing the variance weights, then computing the variance at each grid point

$$V^{(k)} = \sum_{i=1}^{N} \mathcal{N}(|\mathbf{x}_i - \mathbf{x}|, \sigma)(r_i - r^{(k(i))})^2,$$
$$v^{(k)} = \alpha^{(k)} \frac{V^{(k)}}{w^{(k)}} + (1 - \alpha^{(k)}) v_0, \tag{4.76}$$

with v_0 estimated as the average over all variance contributions.

Variations to the Kernel DM+V algorithm have been done to include time dependence [10] and wind information [82].

4.2.1.4 Dynamic Mode Decomposition

In many complex fluid systems, reduced-order modeling has been explored for applications in control and engineering design. Data-driven approaches have been pursued to find effective reduced-order models to represent the system dynamics using only measurement data. Therefore approaches like dynamic mode decomposition (DMD)[87, 65], has shown to be a powerful tool in reducing nonlinear dynamic behavior into a low rank linear approximation. Given a snapshot of the data at timestep k, the states $\mathbf{x}_k$ can be reshaped into a long $nm \times 1$ column vector. By collecting $M + 1$ snapshots the following augmented matrices can be formed,

$$\mathbf{X}_1 = \begin{bmatrix} | & | & & | \\ \mathbf{x}_1 & \mathbf{x}_2 & \cdots & \mathbf{x}_M \\ | & | & & | \end{bmatrix}, \quad \mathbf{X}_2 = \begin{bmatrix} | & | & & | \\ \mathbf{x}_2 & \mathbf{x}_3 & \cdots & \mathbf{x}_{M+1} \\ | & | & & | \end{bmatrix}. \tag{4.77}$$

The idea is to find the optimal local linear approximation of $\mathbf{A}$ such that,

$$\mathbf{X}_2 \approx \mathbf{A}\mathbf{X}_1 \rightarrow, \ \mathbf{A} = \mathbf{X}_2\mathbf{X}_1^{\dagger}, \tag{4.78}$$

where the operator ($\dagger$) represents the Moore-Penrose pseudoinverse. Using the singular value decomposition (SVD) of the augmented matrix $X_1 = U\Sigma V^*$, the r leading eigenvalues can be identified. The best fit reduced order $r \times r$ projection of $\mathbf{A}$ is,

$$\tilde{\mathbf{A}} = \mathbf{U}_r^T \mathbf{X}_2 \mathbf{V}_r \Sigma_r^{-1}. \tag{4.79}$$

Then the eigenvectors $\mathbf{W}$ and eigenvalues Λ of the low rank approximation $\tilde{\mathbf{A}}$ can be used to get the DMD modes of the system,

$$\tilde{\mathbf{A}}\mathbf{W} = \mathbf{W}\Lambda, \tag{4.80}$$

$$\Phi = \mathbf{X}_2\mathbf{V}_r\Sigma_r^{-1}\mathbf{W}. \tag{4.81}$$

The DMD modes can be reconstructed by the columns of Φ give some time step k,

$$\Phi_k = \begin{bmatrix} | & | & & | \\ \phi_1 & \phi_2 & \cdots & \phi_r \\ | & | & & | \end{bmatrix}_k \rightarrow \phi_{1,k} = \begin{bmatrix} \phi_{(1,1)} & \phi_{(1,2)} & \cdots & \phi_{(1,m)} \\ \vdots & \ddots & & \vdots \\ \phi_{(n,1)} & \phi_{(n,2)} & \cdots & \phi_{(n,m)} \end{bmatrix}_{1,k}. \tag{4.82}$$

The number of modes and which frequency of modes to consider can vary from problem to problem.

4.2.2 Multi-sUAS Strategies

4.2.2.1 CVT Coverage Control

The coverage control problem can involve applications such as path planning, obstacle avoidance or deployment of mobile sensors to name a few. Given a domain of interest Ω, the local positions of the sensors $P = (p_1, p_2, ...p_N)$ can be divided into N polytopes $\mathcal{V} = (\mathcal{V}_1, \mathcal{V}_2, ...\mathcal{V}_N)$. The polytopes are determined using centroidal Voronoi tesselations (CVT)[23]. The cost function that controls the CVT desired positions is

$$J(P) = \sum_{i=1}^{N} \int_{\mathcal{V}_i} \rho(q)|q - p_i|^2 dq, \ \text{for} \ q \in \Omega. \tag{4.83}$$

The partial derivative can be used with the mass, m_i, and center of mass, c_i, to identify the critical point,

$$m_i = \int_{\mathcal{V}_i} \rho(q,t)dq, \quad c_i = \frac{1}{m_i}\int_{\mathcal{V}_i} q\rho(q,t)dq, \tag{4.84}$$

$$\frac{\partial J(P)}{\partial p_i} = 2m_i(p_i - c_i)^T. \tag{4.85}$$

This means that $c_i = p_i$ for all $i = 1, 2, ..N$ is a minimizer of the CVT. Treating the relation as a gradient, Llyod's algorithm can be used which is given by

$$\dot{p}_i = -k(p_i - c_i). \tag{4.86}$$

This general algorithm has been used to compute the locations of actuators by computing the CVT using Lloyd's method [25, 26, 24]. This approach is also described in the context of multi-sUAS [55] .

4.3 ASSESSMENT AND SUMMARY OF METHODS

In an attempt to analyze the methods covered in the previous section, we decided to use the following Figure of Merit metrics, such as: required assumptions, sample distance, survey time, complexity, average precision, average accuracy, and average cost. The required assumptions are meant to inform the practitioner so that the best method can be applied to a given problem. For example, if the source location is unknown, the PSG method may not be directly applicable unless a source location estimate is supplied. The sample distance is defined as the distance from the source at which the required method is able to take measurements. The survey time consists of the time required to make a single flux estimate. Understandably, some methods may require multiple flux estimates in order to approximate the emission source to within an acceptable error. Complexity is the measure of how difficult it is to implement any given method. In order to determine a value for complexity, a scheme was developed using figures of merit (FOM) that assigns factors and weights to the metrics (detailed in Table 4.1). Determining the values for these factors was based on loose estimates, inferred from papers found in the literature. Ranges were assigned to the metrics to capture variations in the factors due to either the operators or the equipment being used, and are given in Table 4.2. For example, some setups may use more expensive equipment or more people for the same method, and as a result are reflected in the complexity metric.

Evaluating methane quantification techniques is important and much work has already gone into this topic through controlled release experiments and evaluation

TABLE 4.1 Figures of merit for defining complexity of an estimation method.

FOM	(%)	Low (2.5)	Medium (5)	Med-High (7.5)	High (10)
Operator skill	30	Little	Moderate	Professional	Expert
Number of operators	25	1	2	3	3+
Equipment cost	15	<$10,000	<$50,000	<$100,000	>$100,000
Setup Time	20	<1hr	<4hr	<8hrs	8+hrs
Survey Time	10	<0.5hr	<1hr	<2hrs	2+hrs

TABLE 4.2 Summary of methods and their assumptions, operational details, complexity, cost, average precision, and average accuracy are generalized over implementations found in the literature. The average precision and accuracy are given as unitless values, normalized by the true source rate (estimated source rate). (measured using [1]fixed, [2]foot, [3]vehicle, [4]manned aircraft, or [5]sUAS; cost with $ ≤$10,000, $$ ≤$50,000, $$$ ≤$100,000, $$$$ ≥$100,000; (·) represents precision normalized on estimated source rate average complexity.

Method	Assumptions	Sample Distance	Survey Time	Complexity (1–10)	Avg Precision	Avg Accuracy	Avg Cost
bLS	x_s, horizontally uniform surface source atmosphere in horizontal equilibrium	20–441m[1]	15min–4days[1]	3.8–8.5[1]	±0.16–0.36[1] (±0.07–0.85)[1]	±0.02–0.30[1]	$-$$$[1]
PSG	x_s, steady state source rate, point source, plume evolution via ground-level Gaussian dispersion with no obstructions	441m[1] 18–500m[3] 50m[5]	4days[1] 37–58min[3] 7.22–20min[5]	4.5–8[1] 3.6–6.1[3] 3.3–5.5[5]	(±0.30)[1], ±0.20–0.67[3] (±0.19–0.47)[3] ±0.31[5]	±0.0022–0.43[3] ±0.50[5]	$$–$$$[3]
PSG-RB	PSG assumptions, vertical eddy diffusivity and wind speed approximated by power law scheme	20–200m[3]	6min[3]	3.6–6.4[3]	—	±0[3]	$$–$$$
PSG-CS	PSG assumptions, continuous source emission, constant wind speed, vertical eddy diffusivity and wind speed approximated by power law	18–106m[3]	20min[3]	3.6–5.9[3]	±0.20–0.67[3] (±0.19–0.47)[3]	±0.02–0.26[3]	$$–$$$[3]
NGI	Constant source rate , σ_x and σ_y linear functions of distance to source	50–82.25m[5]	7.35–29.62min[5]	3.3–5.5[5]	±0.21–0.58[5] (±0.06–0.53)[5]	±0.11–0.13[5]	$-$$[5]
MMD	x_s	12–27m[1]	15min[1]	4.5–8.1[1]	±0.06[1]	±0.10[1]	$$$–$$$$[1]
GDT	Near to no meandering, steady state source rate	3–8km[4]	1hr[4]	5.5–7.8[4]	±0.07[4] (±0.08)[4]	±0.13[4]	$$$–$$$$[4]
VFP	GDT assumptions	4.875–10km[4] 19.08–510m[5]	1.5–4.5hr[4] 20–30min[5]	5.8–8.3[4] 3.3– 5.5[5]	(±0.30–0.53)[4] ±0.17–0.37[5] (±0.013–0.62)[5]	±0.10–0.50[4] ±0.03–0.50[5]	$$$–$$$$[4] $-$$[5]
PI-VFP	GDT assumptions	3km[4] 0–6.77m[5]	20–30min[4] 15–20min[5]	5.3–7.5[4] 3.3– 5.5[5]	(±0.34–0.58)[4] ±0.82[5]	±0.27–52[5]	$$$–$$$$[4] $-$$[5]
CFP	GDT assumptions	3–17.84km[4]	1hr[4]	5.5–7.8[4]	—	—	$$$–$$$$[4]
GLM-VFP	GDT assumptions	0.4–2.2km[4]	2.5hr[4]	6–8.3[4]	(±0.21)[4]	—	$$$–$$$[4]
VRPM	GDT assumptions	10–100m[1]	1hr[1]	7.1–8.6[1]	±0.18–0.21[1] (±0.21–0.33)[1]	±0.05–0.43[1]	$-$$$[1]
QOGI	Temperature and pressure of gas at leak location are the same, gas plume length in direction of optical path is small	30m[1]	1min[1]	3–6.3[1]	±0.01–0.02[1] (±0.02–0.03)[1]	±0.20–0.24[1]	$-$$$[1]
TCM	leak plume and tracer plume are well mixed	100–3546m[3]	0.5–2hr[3]	4.6–7.6[3]	±0.06–0.24[3] (±0.06–0.74)[3]	±0.0056–0.17[3]	$$–$$$[3]
EC	Stationarity, fully developed turbulent conditions	25–228m[1]	—	4.5–7.8[1]	(±0.08)[1]	—	$-$$$[1]

frameworks. Examples from controlled release facilities (CRF) consist of but are not limited to the following: In the Joint Urban 2003 study [8, 67], static sensors were distributed in an urban setting to measure the dispersion of tracer particulates; in [78], area-averaged velocity and turbulent kinetic energy profiles are derived from data collected at the Mock Urban Setting Test (MUST); MUST was also evaluated with photo-ionization detectors (PID) [13, 14]; MUST was further simulated using MISKAM 6 [37]; in [77], the WRF model was used to model wind and turbulence inside the Quick Urban and Industrial Complex (QUIC) model for comparing simulated and observed plume transport; a test plan for Jack Rabbit II was developed in [79] which aimed to improve chemical hazard modeling, produce better planning for release incidents, improve emergency response, and improve mitigation measures.

More recently, single-blind tests at the Methane Emission Technology Evaluation Center (METEC) in Fort Collins, Colorado evaluated several types of LDAQ sensing modalities as a part of the Standford/EDF Mobile Monitoring Challenge (MMC) and the Advanced Research Projects Agency-Energy (ARPA-E) MONITOR program (such as by vehicle, plane, and drone – see [81, 18] for more details). In the Standford EDF MMC it was observed that the drone-based technologies performed quite well (e.g. SeekOps) with an $R^2 = 0.42$ [81]. While the results in [81, 18] seem quite promising, there still exist some improvements in precision that can be done. In the ARPA-E MONITOR program, 6 of the 11 participants tested their technologies at the METEC facility in [12] against 6 other industry-based participants. Due to confidentiality agreements at the time of testing, the data gathered from the 12 participants were aggregated to compare the methodologies based on measurement type (handheld, mobile and continuous monitoring). However, to the best of authors knowledge, only four of the MONITOR program participants have published data regarding the METEC tests (see [106, 4, 5, 18]). In the white paper by Bridger Photonics [18], a sUAS-based approach using LiDAR-based sensor has also shown promising results even though the uncertainty is not given. In [103], a RMLD is used on a sUAS with the PI-VFP method. In contrast, [106] utilized a portable TDLAS-based instrument and the PSG method to quantify emissions. Lastly, [4] uses a dual frequency comb spectrometer (from over one kilometer away) with the non-zero minimum bootstrap method (see [5]) and the Gaussian plume model to estimate the source rate. Examples from active operations with comparison to conventional OGI-based methods are conducted in the Alberta Methane Field Challenge (AMFC) [102, 80, 93] which aim to answer the questions: Are leak detection and repair (LDAR) programs effective at reducing methane emissions and can new technologies provide more cost-effective leak detection compared to existing approaches?

In order to compare the performances of the each of the methods to one another, their performance metrics were garnered from different studies where the method was utilized in either a field study or controlled release scenarios and recorded in Table 4.2. Performance values were gathered from the standard deviations of consecutive flux estimates of a single source leak scenario. Accuracy pertains to error of the flux estimate to the known source rate. This information was limited primarily to controlled release scenarios. For each method, performances and details were separated into the broad types of sampling strategies: fixed/static, on foot, mounted on a vehicle, mounted on an aircraft, and mounted on a sUAS. This prevents convolution of performance values between, for example, long aircraft sampling flights at far distances and short sampling flights near the source via sUAS.

4.3.1 Summary of Methods

After analyzing the quantification methods we can separate methods based on whether they have used sUAS or not. In this book we observed that the sUAS-based methods consist of: Near-field Gaussian plume inverse (NGI), vertical flux plane (VFP), and the path-integrated vertical flux plane (PI-VFP). The non-sUAS-based methods consist of: backwards Lagrangian Stochastic (bLS), point source Gaussian (PSG), recursive Bayesian point source Gaussian (PSG-RB), conditionally sampled point source Gaussian (PSG-CS), micrometeorological mass difference (MMD), Gauss divergence theorem (GDT), VFP, PI-VFP, cylindrical flux plane (CFP), general linear model vertical flux plane (GLM-VFP), vertical radial plume mapping (VRPM), quantitative optical gas imaging (QOGI), tracer correlation method (TCM), and Eddy covariance (EC). When comparing their performances in Table 4.2, it can be seen that, when categorizing by means of mobility (i.e. fixed, on-foot, etc.), methods using static sensors show a trend of having higher complexity values while sUAS-based methods display generally lower complexity values. For a subset of the methods, the survey times, sample distances, and average accuracies can be seen in Fig. 4.6. This subset was specifically displayed for these methods had both upper and lower bounds for survey times and sample distances along with accuracy data which allowed for the plotting of these quantities for each method in the form of ellipses on a log-log plot. When analyzing this plot, it can be seen that the sUAS-based methods are generally lower in sample distances and survey times as opposed to manned aircraft-based methods (being the one of the highest in both). The bLS and TCM methods are shown to have the best average accuracy with several sUAS and mobile methods close in accuracy. The long sample times of bLS method are due to the values reported in [91], and it is possible that these values don't reflect typical bLS sample times. The advantages and disadvantages of each of the methods can be seen in Table 4.3 along with what typical application fields that they were applied in.

The final ranking of the methods depends heavily on the desired application, which also depends on factors such as sample distance, sample time, and desired accuracy. For that reason, it is difficult to rank the methods in general. Thus, we provide a ranking of the methods in terms of complexity (outlined in Table 4.1) with highlights from the precision and cost in Fig. 4.7. The results indicate that the simplest methods, in terms of complexity, are sUAS-based NGI[5] and VFP[5] as well as fixed QOGI[1]. The most complex methods include bLS[1] and manned aircraft-based approaches. In terms of precision, bLS[1], NGI[5], GDT[4], VFP[5], QOGI[1], TCM[3], and EC[1] tend to be the best. Thus, for sUAS-based methods, NGI[5] and VFP[5], are the most promising approaches. Additionally, the GDT[4], TCM[3], and EC[1] approaches can be treated as candidate methods for future implementation using sUAS.

TABLE 4.3 Summary of method advantages and disadvantages along with fields of application.

Method	Application	Advantages	Disadvantages
bLS	General, Oil and Gas, Agriculture	Able to quantify area source and point source emissions	Sensor is fixed, multiple measurements, negatively impacted by obstacles
PSG	Biogas, Oil and Gas	No tracer required	Negatively impacted by obstacles and low wind speeds due to plume advection, mobile sensors limited to roads
PSG-RB	Oil and Gas	No tracer required, accurate source quantification in open environments	*PSG limitations*
PSG-CS	Oil and Gas	No tracer required	*PSG limitations*
NGI	General, Oil and Gas	Assumes near-field plume turbulence and wind meandering	Near-field (100m), prop-wash interference if unaccounted for
MMD	Agriculture	Able to give instantaneous flux estimates	Fixed sensors
GDT	Regional	Capable of giving emission quantifications of large areas	Sample around closed volume typically large areas, unable to capture instantaneous methane flux
VFP	General, Oil and Gas, Landfill	Does not require exact source location, ease of mobility	Stable atmospheric conditions, unable to capture instantaneous methane flux
PI-VFP	General, Oil and Gas	*VFP advantages*	*VFP limitations*
CFP	General, Urban	capable of giving emission quantifications of large areas	*GDT limitations*
GLM-VFP	Landfill	*VFP advantages*	*VFP limitations*
VRPM	Landfill	Able to give instantaneous flux estimates	Fixed sensors
QOGI	Biogas	able to give instantaneous flux estimates	Gas velocity determined via gas camera – velocity component parallel to image plane, can be difficult to process images
TCM	General, Oil and Gas, Landfill, Urban	Does not rely on meteorological measurements	Mobile sensors limited to roads, application difficulty due to outside methane source interference
EC	General, Oil and Gas	Can capture emission variations due to long time series	Fixed sensor(s), stable atmospheric conditions, sensitive to time of day, typically requires long sampling times

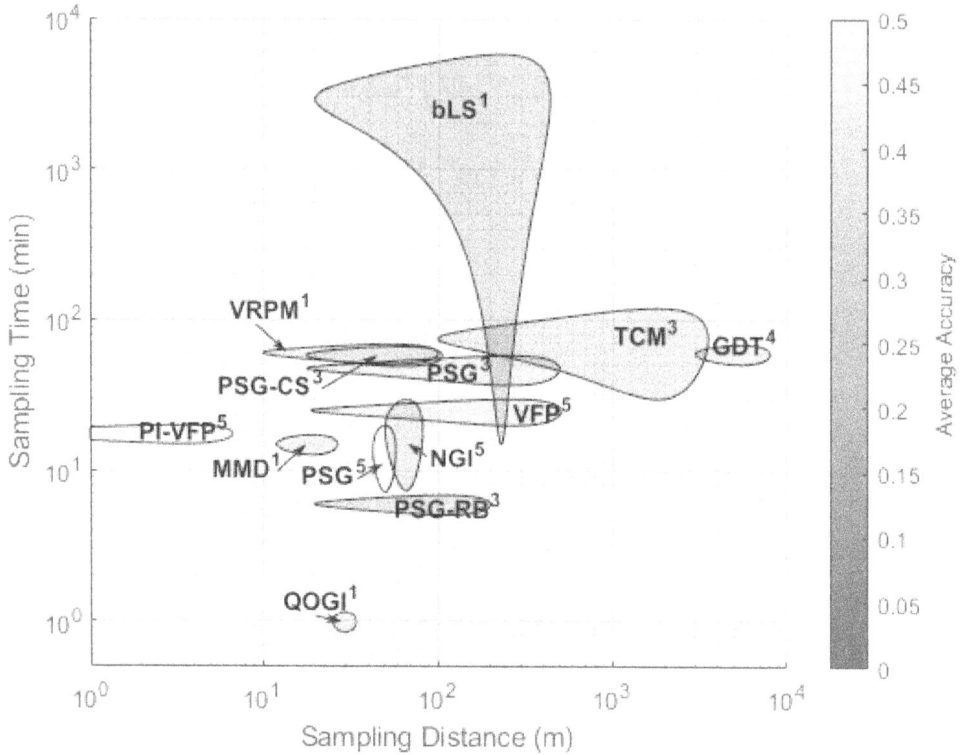

FIGURE 4.6 Diagram of summary of methods (based on Table 4.2) showing relationship between typical survey time versus sample distance and there associated normalized accuracy, where lower values represent more accurate measurements. (measured using [1]fixed, [2]foot, [3]vehicle, [4]manned aircraft, or [5]sUAS) [57].

Pause and Reflect

As sensors become smaller, lighter, and cheaper in the future, what kinds of quantification approaches will become possible for sUAS?

4.4 CHAPTER SUMMARY

In this chapter we overview the types of quantification methodologies that are within the literature, strategies that can be applied to sUAS-based quantification techniques, and an assessment of the methods covered in terms of cost, complexity, and precision. The quantification methods covered in this chapter were meant to be relevant to the problem of estimating emissions from remotely measured data but also that they can be implemented, in theory, with sUAS. For example, OGI cameras have become small enough and sensitive enough to be utilized in sUAS-based detection and quantification tasks. Additionally, as edge computation technologies improve, the possibility to

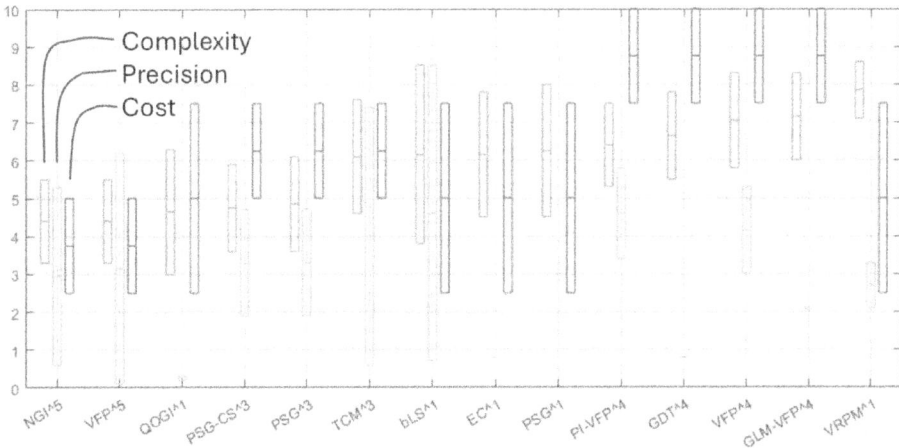

FIGURE 4.7 Diagram of the complexity ranking of the methods (based on Table 4.1), showing the relationship between the method complexity, precision, and cost. The precision is normalized on the source estimate multiplied by 10 and the cost is ranked from 0 to 10. (measured using [1]fixed, [2]foot, [3]vehicle, [4]manned aircraft, or [5]sUAS) [57]

undergo real-time optimization or conduct real-time simulation techniques become increasingly probable. For example, with advances in machine learning, plume extraction from OGI cameras can be utilized to quantify emissions from methods like the integrated mass enhancement. In practice, the near-field Gaussian plume inversion (NGI) method and the mass-balance methods (e.g. vertical flux plane, cylindrical flux plane, path-integrate, and GDT) are being applied for many applications but variations such as the pumped TDLAS (as seen in Chapter 2), which are unique for surface based fluxes (e.g. landfills), are gaining in popularity.

Bibliography

[1] Adel A Abdel-Rahman. On the atmospheric dispersion and Gaussian plume model. In *WWAI'08: Proc. of the 2nd International Conference on Waste Management, Water Pollution, Air Pollution, Indoor Climate*, pages 31–39, 2008.

[2] Adil Shah. Supplement to A Near-Field Gaussian Plume Inversion Flux Quantification Method, Suitable For Unmanned Aerial Vehicle Sampling. *Atmosphere*, 2020.

[3] John D Albertson, Tierney Harvey, Greg Foderaro, Pingping Zhu, Xiaochi Zhou, Silvia Ferrari, M Shahrooz Amin, Mark Modrak, Halley Brantley, and Eben D Thoma. A mobile sensing approach for regional surveillance of fugitive methane emissions in oil and gas production. *Environmental Science & Technology*, 50(5):2487–2497, 2016.

[4] Caroline B Alden, Sean C Coburn, Robert J Wright, Esther Baumann, Kevin Cossel, Edgar Perez, Eli Hoenig, Kuldeep Prasad, Ian Coddington, and Gregory B Rieker. Single-blind quantification of natural gas leaks from 1 km distance using frequency combs. *Environmental science & Technology*, 53(5):2908–2917, 2019.

[5] Caroline B Alden, Subhomoy Ghosh, Sean Coburn, Colm Sweeney, Anna Karion, Robert Wright, Ian Coddington, Gregory B Rieker, and Kuldeep Prasad. Bootstrap inversion technique for atmospheric trace gas source detection and quantification using long open-path laser measurements. *Atmospheric Measurement Techniques*, 11(3):1565–1582, 2018.

[6] David Allen, Shannon Stokes, Erin Tullos, Brendan Smith, Scott Herndon, and Bradley Flowers. Field trial of methane emission quantification technologies. In *Proc. of the SPE Annual Technical Conference and Exhibition*. OnePetro, 2020.

[7] Grant Allen, Peter Hollingsworth, Khristopher Kabbabe, Joseph R Pitt, Mohammed I Mead, Samuel Illingworth, Gareth Roberts, Mark Bourn, Dudley E Shallcross, and Carl J Percival. The development and trial of an unmanned aerial system for the measurement of methane flux from landfill and greenhouse gas emission hotspots. *Waste Management*, 87:883–892, 2019.

[8] K Jerry Allwine and Julia E Flaherty. Joint Urban 2003: Study overview and instrument locations. Technical report, Pacific Northwest National Lab.(PNNL), Richland, WA (United States), 2006.

[9] C Arndt, AB Leytem, AN Hristov, D Zavala-Araiza, JP Cativiela, S Conley, C Daube, Ian Faloona, and SC Herndon. Short-term methane emissions from 2 dairy farms in California estimated by different measurement techniques and US Environmental Protection Agency inventory methodology: A case study. *Journal of Dairy Science*, 101(12):11461–11479, 2018.

[10] Sahar Asadi, Han Fan, Victor Hernandez Bennetts, and Achim J Lilienthal. Time-dependent gas distribution modelling. *Robotics and Autonomous Systems*, 96:157–170, 2017.

[11] Antoine Babilotte. Field comparison of methods for assessment of methane fugitive emissions from landfills. *Environmental Research and Education Foundation (EREF)*, 2011.

[12] Clay S Bell, Timothy Vaughn, Daniel Zimmerle, Detlev Helmig, and Brian Lamb. Evaluation of next generation emission measurement technologies under repeatable test protocols. *Elementa: Science of the Anthropocene*, 8, 2020.

[13] C. A. Biltoft. Customer report for mock urban setting test. *DPG Document*, (8-CO):160–000, 2001.

[14] C. A. Biltoft and E Yee. Overview of the Mock Urban Setting Test (MUST) C.A. *Engineering*, 2001.

[15] M Bourn, G Allen, P Hollingsworth, K Kababbe, P I Williams, H Ricketts, J R Pitt, and A Shah. The development of an unmanned aerial system for the measurement of methane emissions from landfill. *Sixteenth International Waste Management and Landfill Symposium*, (October 2017), 2018.

[16] Halley L Brantley, Eben D Thoma, William C Squier, Birnur B Guven, and David Lyon. Assessment of methane emissions from oil and gas production pads using mobile measurements. *Environmental Science & Technology*, 48(24):14508–14515, 2014.

[17] Timo Rolf Bretschneider and Karan Shetti. UAV-based gas pipeline leak detection. *35th Asian Conference on Remote Sensing 2014, ACRS 2014: Sensing for Reintegration of Societies*, (April), 2014.

[18] Bridger Photonics. Gas mapping lidar™ METEC round 1 results.

[19] Gary A Briggs. Plume rise: A critical survey. Technical report, Air Resources Atmospheric Turbulence and Diffusion Lab., Oak Ridge, Tenn., 1969.

[20] George Burba. *Eddy Covariance Method for Scientific, Industrial, Agricultural and Regulatory Applications: A Field Book on Measuring Ecosystem Gas Exchange and Areal Emission Rates*. LI-Cor Biosciences, 2013.

[21] Maria Obiminda L Cambaliza, Jean E Bogner, Roger B Green, Paul B Shepson, Tierney A Harvey, Kurt A Spokas, Brian H Stirm, Margaret Corcoran, Detlev Helmig, and Armin Wisthaler. Field measurements and modeling to resolve m^2 to km^2 CH$_4$ emissions for a complex urban source: An Indiana landfill study. *Elementa: Science of the Anthropocene*, 5, 2017.

[22] MOL Cambaliza, PB Shepson, DR Caulton, B Stirm, D Samarov, KR Gurney, J Turnbull, KJ Davis, A Possolo, A Karion, et al. Assessment of uncertainties of an aircraft-based mass balance approach for quantifying urban greenhouse gas emissions. *Atmospheric Chemistry and Physics*, 14(17):9029–9050, 2014.

[23] Jianxiong Cao, YangQuan Chen, and Changpin Li. Multi-UAV-based optimal crop-dusting of anomalously diffusing infestation of crops. In *Proc. of the 2015 American Control Conference (ACC)*, pages 1278–1283. IEEE, 2015.

[24] YangQuan Chen, Zhongmin Wang, and Jinsong Liang. Actuation scheduling in mobile actuator networks for spatial-temporal feedback control of a diffusion process with dynamic obstacle avoidance. In *IEEE International Conference Mechatronics and Automation, 2005*, volume 2, pages 752–757 Vol. 2, 2005.

[25] YangQuan Chen, Zhongmin Wang, and Jinsong Liang. Optimal dynamic actuator location in distributed feedback control of a diffusion process. *International Journal of Sensor Networks*, 2(3-4):169–178, 2007.

[26] YangQuan Chen, Zhongmin Wang, and K.L. Moore. Optimal spraying control of a diffusion process using mobile actuator networks with fractional potential field based dynamic obstacle avoidance. In *Proc. of the 2006 IEEE International Conference on Networking, Sensing and Control*, pages 107–112, 2006.

[27] Stephen Conley, Ian Faloona, Shobhit Mehrotra, Maxime Suard, Donald H Lenschow, Colm Sweeney, Scott Herndon, Stefan Schwietzke, Gabrielle Pétron, Justin Pifer, et al. Application of Gauss's theorem to quantify localized surface emissions from airborne measurements of wind and trace gases. *Atmospheric Measurement Techniques*, 10(9):3345–3358, 2017.

[28] Noel Cressie. Kriging nonstationary data. *Journal of the American Statistical Association*, 81(395):625–634, 1986.

[29] PM Czepiel, B Mosher, RC Harriss, JH Shorter, JB McManus, CE Kolb, E Allwine, and BK Lamb. Landfill methane emissions measured by enclosure and atmospheric tracer methods. *Journal of Geophysical Research: Atmospheres*, 101(D11):16711–16719, 1996.

[30] Florentino B De la Cruz, Roger B Green, Gary R Hater, Jeffrey P Chanton, Eben D Thoma, Tierney A Harvey, and Morton A Barlaz. Comparison of field measurements to methane emissions models at a new landfill. *Environmental Science & Technology*, 50(17):9432–9441, 2016.

[31] OT Denmead, LA Harper, JR Freney, DWT Griffith, R Leuning, and RR Sharpe. A mass balance method for non-intrusive measurements of surface-air trace gas exchange. *Atmospheric Environment*, 32(21):3679–3688, 1998.

[32] John   D'Errico.   fminsearchbnd,   fminsearchcon. https://www.mathworks.com/matlabcentral/fileexchange/8277-fminsearchbnd-fminsearchcon, Feb 2022.

[33] Sören Dierks and Andreas Kroll. Quantification of methane gas leakages using remote sensing and sensor data fusion. In *Proc. of the 2017 IEEE Sensors Applications Symposium (SAS)*, pages 1–6. IEEE, 2017.

[34] Richard M Eckman. Re-examination of empirically derived formulas for horizontal diffusion from surface sources. *Atmospheric Environment*, 28(2):265–272, 1994.

[35] Rachel Edie, Anna M Robertson, Robert A Field, Jeffrey Soltis, Dustin A Snare, Daniel Zimmerle, Clay S Bell, Timothy L Vaughn, and Shane M Murphy. Constraining the accuracy of flux estimates using OTM 33A. *Atmospheric Measurement Techniques*, 13(1):341–353, 2020.

[36] Rachel Edie, Anna M Robertson, Jeffrey Soltis, Robert A Field, Dustin Snare, Matthew D Burkhart, and Shane M Murphy. Off-site flux estimates of volatile organic compounds from oil and gas production facilities using fast-response instrumentation. *Environmental Science & Technology*, 54(3):1385–1394, 2019.

[37] Joachim Eichhorn and Márton Balczó. Flow and dispersal simulations of the mock urban setting test. *Hrvatski meteorološki časopis*, 43(43/1):67–72, 2008.

[38] Benjamin Fasoli, John C Lin, David R Bowling, Logan Mitchell, and Daniel Mendoza. Simulating atmospheric tracer concentrations for spatially distributed receptors: updates to the stochastic time-inverted lagrangian transport model's R interface (STILT-R version 2). *Geoscientific Model Development*, 11(7):2813–2824, 2018.

[39] Thomas K Flesch, John D Wilson, and Eugene Yee. Backward-time lagrangian stochastic dispersion models and their application to estimate gaseous emissions. *Journal of Applied Meteorology and Climatology*, 34(6):1320–1332, 1995.

[40] Markus D Foote, Philip E Dennison, Andrew K Thorpe, David R Thompson, Siraput Jongaramrungruang, Christian Frankenberg, and Sarang C Joshi. Fast and accurate retrieval of methane concentration from imaging spectrometer data using sparsity prior. *IEEE Transactions on Geoscience and Remote Sensing*, 58(9):6480–6492, 2020.

[41] Tierney A Foster-Wittig, Eben D Thoma, and John D Albertson. Estimation of point source fugitive emission rates from a single sensor time series: A conditionally-sampled Gaussian plume reconstruction. *Atmospheric Environment*, 115:101–109, 2015.

[42] Tierney A Foster-Wittig, Eben D Thoma, Roger B Green, Gary R Hater, Nathan D Swan, and Jeffrey P Chanton. Development of a mobile tracer correlation method for assessment of air emissions from landfills and other area sources. *Atmospheric Environment*, 102:323–330, 2015.

[43] James L France, Prudence Bateson, Pamela Dominutti, Grant Allen, Stephen Andrews, Stephane Bauguitte, Max Coleman, Tom Lachlan-Cope, Rebecca E Fisher, Langwen Huang, et al. Facility level measurement of offshore oil and gas installations from a medium-sized airborne platform: method development for quantification and source identification of methane emissions. *Atmospheric Measurement Techniques*, 14(1):71–88, 2021.

[44] Christian Frankenberg, Andrew K Thorpe, David R Thompson, Glynn Hulley, Eric Adam Kort, Nick Vance, Jakob Borchardt, Thomas Krings, Konstantin Gerilowski, Colm Sweeney, et al. Airborne methane remote measurements reveal heavy-tail flux distribution in four corners region. *Proc. of the National Academy of Sciences*, 113(35):9734–9739, 2016.

[45] Michael B. Frish. Monitoring fugitive methane emissions utilizing advanced small unmanned aerial sensor technology. 2016. http://www.psicorp.com/sites/psicorp.com/files/articles/SR-2018-3.pdf.

[46] Montserrat Fuentes. A high frequency kriging approach for non-stationary environmental processes. *Environmetrics: The Official Journal of the International Environmetrics Society*, 12(5):469–483, 2001.

[47] Zhiling Gao, Raymond L Desjardins, and Thomas K Flesch. Comparison of a simplified micrometeorological mass difference technique and an inverse dispersion technique for estimating methane emissions from small area sources. *Agricultural and Forest Meteorology*, 149(5):891–898, 2009.

[48] D. Gasbarra, P. Toscano, D. Famulari, S. Finardi, P. Di Tommasi, A. Zaldei, P. Carlucci, E. Magliulo, and B. Gioli. Locating and quantifying multiple landfills methane emissions using aircraft data. *Environmental Pollution*, 254:112987, 2019.

[49] Mathias Göckede, Corinna Rebmann, and Thomas Foken. A combination of quality assessment tools for eddy covariance measurements with footprint modelling for the characterisation of complex sites. *Agricultural and Forest Meteorology*, 127(3-4):175–188, 2004.

[50] Jake R Gemerek, Silvia Ferrari, and John D Albertson. Fugitive gas emission rate estimation using multiple heterogeneous mobile sensors. In *Proc. of the 2017 ISOCS/IEEE International Symposium on Olfaction and Electronic Nose (ISOEN)*, pages 1–3. IEEE, 2017.

[51] Pierre Goovaerts. Kriging and semivariogram deconvolution in the presence of irregular geographical units. *Mathematical Geosciences*, 40(1):101–128, 2008.

[52] Md.Hasibul Hasan, Poyu Zhang, Jiannan Chen, Guoliang Shi, Tarek Abichou, and Haofei Yu. Exploring uncertainties in the integrated mass enhancement method for remote sensing retrievals of methane emissions. *Waste Management*, 200:114759, 2025.

[53] Bruce B Hicks and Dennis D Baldocchi. Measurement of fluxes over land: Capabilities, origins, and remaining challenges. *Boundary-Layer Meteorology*, 177:365–394, 2020.

[54] Derek Hollenbeck and YangQuan Chen. Characterization of ground-to-air emissions with sUAS using a digital twin framework. In *Proc. of the 2020 International Conference on Unmanned Aircraft Systems (ICUAS)*, pages 1162–1166. IEEE, 2020.

[55] Derek Hollenbeck and YangQuan Chen. Mutli-UAV method for continuous source rate estimation of fugitive gas emissions from a point source. In *Proc. of the 2021 International Conference on Unmanned Aircraft Systems (ICUAS)*. IEEE, 2021.

[56] Derek Hollenbeck, Kristen Manies, YangQuan Chen, Dennis Baldocchi, Eugenie Euskirchen, and Lance Christensen. Evaluating a UAV-based mobile sensing system designed to quantify ecosystem-based methane. *Earth and Space Science Open Archive*, page 15, 2021.

[57] Derek Hollenbeck, Demitrius Zulevic, and Yangquan Chen. Advanced leak detection and quantification of methane emissions using sUAS. *Drones*, 5(4):117, 2021.

[58] Derek Hollenbeck, Demitrius Zulevic, and YangQuan Chen. A modified near-field Gaussian plume inversion method using multi-sUAS for emission quantification. In *2022 International Conference on Unmanned Aircraft Systems (ICUAS)*, pages 1620–1625. IEEE, 2022.

[59] C Hunter. A recommended Pasquill-Gifford stability classification method for safety basis atmospheric dispersion modeling at SRS. Technical report, Savannah River Site (SRS), 2012.

[60] Michael Hutchinson, Cunjia Liu, and Wen-Hua Chen. Source term estimation of a hazardous airborne release using an unmanned aerial vehicle. *Journal of Field Robotics*, 36(4):797–817, 2019.

[61] Paul Imhoff, Ramin Yazdani, Byunghyun Han, Changen Mei, and Don Augenstein. Quantifying capture efficiency of gas collection wells with gas tracers. *Waste Management*, 43:319–327, 2015.

[62] Paul T Imhoff, Fotini K Chow, Diane Taylor, and Madjid Delkash. Assessing Accuracy of Tracer Dilution Measurements of Methane Emissions from Landfills with Wind Modeling. Technical report, Environmental Research and Education Foundation, 2018.

[63] Yuhan Jiang, Lu Zhang, Xingying Zhang, and Xifeng Cao. Methane retrieval algorithms based on satellite: A review. *Atmosphere*, 15(4):449, 2024.

[64] Taylor S Jones, Jonathan E Franklin, Jia Chen, Florian Dietrich, Kristian D Hajny, Johannes C Paetzold, Adrian Wenzel, Conor Gately, Elaine Gottlieb, Harrison Parker, et al. Assessing urban methane emissions using column-observing portable Fourier transform infrared (FTIR) spectrometers and a novel Bayesian inversion framework. *Atmospheric Chemistry and Physics*, 21(17):13131–13147, 2021.

[65] J Nathan Kutz, Steven L Brunton, Bingni W Brunton, and Joshua L Proctor. *Dynamic Mode Decomposition: Data-Driven Modeling of Complex Systems*. SIAM, 2016.

[66] Brian K Lamb, J Barry McManus, Joanne H Shorter, Charles E Kolb, Byard Mosher, Robert C Harriss, Eugene Allwine, Denise Blaha, Touche Howard, Alex Guenther, et al. Development of atmospheric tracer methods to measure methane emissions from natural gas facilities and urban areas. *Environmental Science & Technology*, 29(6):1468–1479, 1995.

[67] MJ Leach. Final report for the Joint Urban 2003 atmospheric dispersion study in Oklahoma City: Lawrence Livermore National Laboratory participation.

Technical report, Lawrence Livermore National Lab.(LLNL), Livermore, CA (United States), 2005.

[68] Henning Lebrenz and András Bárdossy. Geostatistical interpolation by quantile kriging. *Hydrology and Earth System Sciences*, 23(3):1633–1648, 2019.

[69] Ira Leifer, Christopher Melton, Marc L Fischer, Matthew Fladeland, Jason Frash, Warren Gore, Laura T Iraci, Josette E Marrero, Ju-Mee Ryoo, Tomoaki Tanaka, et al. Atmospheric characterization through fused mobile airborne and surface in situ surveys: methane emissions quantification from a producing oil field. *Atmospheric Measurement Techniques*, 11(3):1689–1705, 2018.

[70] JC Lin, Christoph Gerbig, SC Wofsy, AE Andrews, BC Daube, KJ Davis, and CA Grainger. A near-field tool for simulating the upstream influence of atmospheric observations: The stochastic time-inverted Lagrangian transport (STILT) model. *Journal of Geophysical Research: Atmospheres*, 108(D16), 2003.

[71] George Y Lu and David W Wong. An adaptive inverse-distance weighting spatial interpolation technique. *Computers & Geosciences*, 34(9):1044–1055, 2008.

[72] Alessandro Mazzella and Antonio Mazzella. The importance of the model choice for experimental semivariogram modeling and its consequence in evaluation process. *Journal of Engineering*, 2013, 2013.

[73] Jacob Mønster, Peter Kjeldsen, and Charlotte Scheutz. Methodologies for measuring fugitive methane emissions from landfills–a review. *Waste Management*, 87:835–859, 2019.

[74] Jacob G Mønster, Jerker Samuelsson, Peter Kjeldsen, Chris W Rella, and Charlotte Scheutz. Quantifying methane emission from fugitive sources by combining tracer release and downwind measurements–a sensitivity analysis based on multiple field surveys. *Waste Management*, 34(8):1416–1428, 2014.

[75] Randulph P Morales, Jonas Ravelid, Killian P Brennan, Béla Tuzson, Lukas Emmenegger, and Dominik Brunner. Estimating local methane sources from drone-based laser spectrometer measurements by mass-balance method. In *EGU General Assembly Conference Abstracts*, page 14778, 2020.

[76] Thomas Nehrkorn, Janusz Eluszkiewicz, Steven C Wofsy, John C Lin, Christoph Gerbig, Marcos Longo, and Saulo Freitas. Coupled weather research and forecasting–stochastic time-inverted Lagrangian transport (WRF–STILT) model. *Meteorology and Atmospheric Physics*, 107:51–64, 2010.

[77] Matthew A Nelson, Michael J Brown, Scot A Halverson, Paul E Bieringer, Andrew Annunzio, George Bieberbach, and Scott Meech. A case study of the Weather Research and Forecasting model applied to the Joint Urban 2003

tracer field experiment. part 2: Gas tracer dispersion. *Boundary-Layer Meteorology*, 161(3):461–490, 2016.

[78] Matthew A Nelson, MJ Brown, ER Pardyjak, and JC Klewicki. Area-averaged profiles over the Mock Urban Setting Test array. Technical report, Los Alamos National Laboratory, 2004.

[79] Damon K Nicholson, Allison Hedrick, Petr Serguievski, and Allyssa A Martinez. Detailed test plan for Jack Rabbit (JR) II. Technical report, West Desert Test Center Dugway Proving Ground UT, 2015.

[80] Arvind P Ravikumar, Brenna Barlow, Jiayang Wang, and Devyani Singh. Results from the Alberta Methane Measurement Campaigns: New insights into oil and gas methane mitigation policy. In *AGU Fall Meeting Abstracts*, volume 2019, pages A41D–08, 2019.

[81] Arvind P Ravikumar, Sindhu Sreedhara, Jingfan Wang, Jacob Englander, Daniel Roda-Stuart, Clay Bell, Daniel Zimmerle, David Lyon, Isabel Mogstad, Ben Ratner, et al. Single-blind inter-comparison of methane detection technologies–results from the Stanford/EDF Mobile Monitoring Challenge. *Elementa: Science of the Anthropocene*, 7, 2019.

[82] Matteo Reggente and Achim J Lilienthal. The 3D-kernel DM+V/W algorithm: Using wind information in three dimensional gas distribution modelling with a mobile robot. In *SENSORS, 2010 IEEE*, pages 999–1004. IEEE, 2010.

[83] Maximilian Reuter, Heinrich Bovensmann, Michael Buchwitz, Jakob Borchardt, Sven Krautwurst, Konstantin Gerilowski, Matthias Lindauer, Dagmar Kubistin, and John P. Burrows. Development of a small unmanned aircraft system to derive CO_2 emissions of anthropogenic point sources. *Atmospheric Measurement Techniques*, 14(1):153–172, 2021.

[84] JR Roscioli, TI Yacovitch, C Floerchinger, AL Mitchell, DS Tkacik, R Subramanian, DM Martinez, TL Vaughn, L Williams, D Zimmerle, et al. Measurements of methane emissions from natural gas gathering facilities and processing plants: measurement methods. *Atmospheric Measurement Techniques*, 8(5):2017–2035, 2015.

[85] Ju Mee Ryoo, Laura T. Iraci, Tomoaki Tanaka, Josette E. Marrero, Emma L. Yates, Inez Fung, Anna M. Michalak, Jovan Tadić, Warren Gore, T. Paul Bui, Jonathan M. Dean-Day, and Cecilia S. Chang. Quantification of CO_2 and CH_4 emissions over Sacramento, California, based on divergence theorem using aircraft measurements. *Atmospheric Measurement Techniques*, 12(5):2949–2966, 2019.

[86] Rebecca Del Papa Moreira Scafutto and Carlos Roberto de Souza Filho. Detection of methane plumes using airborne midwave infrared (3-5 μm) hyperspectral data. *Remote Sensing*, 10(8):1–16, 2018.

[87] Peter J Schmid. Dynamic mode decomposition of numerical and experimental data. *Journal of Fluid Mechanics*, 656:5–28, 2010.

[88] Adil Shah, Grant Allen, Joseph R Pitt, Hugo Ricketts, Paul I Williams, Jonathan Helmore, Andrew Finlayson, Rod Robinson, Khristopher Kabbabe, Peter Hollingsworth, et al. A near-field Gaussian plume inversion flux quantification method, applied to unmanned aerial vehicle sampling. *Atmosphere*, 10(7):396, 2019.

[89] Adil Shah, Grant Allen, Hugo Ricketts, Joseph Pitt, and Paul Williams. Methane flux quantification from lactating cattle using unmanned aerial vehicles. *EGU General Assembly*, 20:7655, 2018.

[90] Adil A Shah. Methane Flux Quantification Using Unmanned Aerial Vehicles. *Diss., University of Manchester*, 2020.

[91] Jacob T Shaw, Grant Allen, Joseph Pitt, Adil Shah, Shona Wilde, Laurence Stamford, Zhaoyang Fan, Hugo Ricketts, Paul I Williams, Prudence Bateson, et al. Methane flux from flowback operations at a shale gas site. *Journal of the Air & Waste Management Association*, 70(12):1324–1339, 2020.

[92] Donald Shepard. A two-dimensional interpolation function for irregularly-spaced data. In *Proceedings of the 1968 23rd ACM national conference*, pages 517–524, 1968.

[93] Devyani Singh, Brenna Barlow, Chris Hugenholtz, Wes Funk, Cooper Robinson, and Arvind P Ravikumar. Field performance of new methane detection technologies: Results from the Alberta Methane Field Challenge. 2021.

[94] Giovanni Tanda, Marco Balsi, Paolo Fallavollita, and Valter Chiarabini. A UAV-based thermal-imaging approach for the monitoring of urban landfills. *Inventions*, 5(4):1–13, 2020.

[95] E. Thoma and B. Squier. Draft Other Test Method 33A: Geospatial measurement of air pollution, remote emissions quantification - direct assessment (GMAP-REQ-DA). *Environmental Protection Agency (EPA)*, 2014. https://www3.epa.gov/ttnemc01/prelim/otm33a.pdf.

[96] AK Thorpe, C Frankenberg, and DA Roberts. Retrieval techniques for airborne imaging of methane concentrations using high spatial and moderate spectral resolution: Application to AVIRIS. *Atmospheric Measurement Techniques*, 7(2):491–506, 2014.

[97] Andrew K Thorpe, Christian Frankenberg, David R Thompson, Riley M Duren, Andrew D Aubrey, Brian D Bue, Robert O Green, Konstantin Gerilowski, Thomas Krings, Jakob Borchardt, et al. Airborne DOAS retrievals of methane, carbon dioxide, and water vapor concentrations at high spatial resolution: application to AVIRIS-NG. *Atmospheric Measurement Techniques*, 10(10), 2017.

[98] AP Van Ulden. Simple estimates for vertical diffusion from sources near the ground. *Atmospheric Environment*, 12(11):2125–2129, 1978.

[99] Hans Wackernagel. Ordinary Kriging. In *Multivariate Geostatistics*, pages 79–88. Springer, 2003.

[100] M Whiticar, L Christensen, C Salas, and P Reece. GHGMap: novel approach for aerial measurements of greenhouse gas emissions British Columbia. *Geoscience BC Summary of Activities 2017: Energy, Geoscience BC, Report 2018-4*, pages 1–10, 2018.

[101] M Whiticar, L Christensen, C Salas, and P Reece. GHGMap: Detection of fugitive methane leaks from natural gas pipelines British Columbia and Alberta. *Geoscience BC Summary of Activities 2018: Energy and Water, Geoscience BC, Report 2019-2*, pages 67–76, 2019.

[102] M Whiticar, D Hollenbeck, B Billwiller, C Salas, and L.E Christensen. Application of the BC GHGMapper™ platform for the Alberta Methane Field Challenge (AMFC). *Geoscience BC Summary of Activities 2019: Energy and Water, Geoscience BC, Report 2020-02*, pages 87–102, 2020.

[103] Shuting Yang, Robert W Talbot, Michael B Frish, Levi M Golston, Nicholas F Aubut, Mark A Zondlo, Christopher Gretencord, and James McSpiritt. Natural gas fugitive leak detection using an unmanned aerial vehicle: Measurement system description and mass balance approach. *Atmosphere*, 9(10):383, 2018.

[104] Eugene Yee. Probability theory as logic: data assimilation for multiple source reconstruction. *Pure and Applied Geophysics*, 169(3):499–517, 2012.

[105] Eugene Yee and Thomas K Flesch. Inference of emission rates from multiple sources using Bayesian probability theory. *Journal of Environmental Monitoring*, 12(3):622–634, 2010.

[106] Eric J Zhang, Chu C Teng, Theodore G van Kessel, Levente Klein, Ramachandran Muralidhar, Gerard Wysocki, and William MJ Green. Field deployment of a portable optical spectrometer for methane fugitive emissions monitoring on oil and gas well pads. *Sensors*, 19(12):2707, 2019.

[107] Xiaochi Zhou, Seungju Yoon, Steve Mara, Matthias Falk, Toshihiro Kuwayama, Travis Tran, Lucy Cheadle, Jim Nyarady, Bart Croes, Elizabeth Scheehle, et al. Mobile sampling of methane emissions from natural gas well pads in California. *Atmospheric Environment*, 244:117930, 2021.

CHAPTER 5

Case Studies: LDAQ with sUAS

IN the beginning of this book we learned the importance of methane and how to sense it. In the previous three chapters, we learned about how to detect methane emissions from sUAS (including sensor integration considerations and how to deploy the aircraft), how to localize an emission source (given trajectory of concentration and wind data), and finally how to quantify emission sources using simulation-, optimization-, or mass-balance-based methodologies. The consortium of emission leak detection, localization, and quantification methods explored thus far, make up examples of what we would call leak detection and quantification (LDAQ) methodologies. In theory, we make assumptions to simplify the mathematics or the simulation complexities to extract important emission parameters (e.g. emission source rate, location, etc.). In practice, these theoretical assumptions of the applied methodologies are challenged through field studies. As a result, the methodological performance can ultimately be validated. In this chapter, we will highlight some case study examples in each of these three categories (leak detection, leak localization, and leak quantification).

5.1 LEAK DETECTION AND LOCALIZATION

During the early works of sensor placement on-board the sUAS for methane detection, controlled release studies were undertaken to understand how the integrated system performs in the detection of methane and how that performance may change in different environments. In this case study, we will look at two types of environments, rural and urban. For example, the complexity differences between rural and urban environments vary drastically in the atmospheric stability (e.g. due to changes in surface roughness), wind patterns (e.g. open and flat, versus cluttered with buildings), or other disturbances (e.g. cars, trees, etc.). Once the emission source can be sufficiently detected within some confidence interval, the next operational step is to focus on localizing the source. The performance, as understood in Chapter 3, depends on the choice of method, and how the sUAS measurement systems are deployed. Heuristical approaches are often the easiest and first choice to implement – due to the simplicity.

DOI: 10.1201/9781003669470-5

These can be generalized as a 'rule of thumb' approach, and can be quite useful for localizing or even just guiding the searcher toward the source. In this section we will focus on using back trajectory approach. However, as we will see in this case study, this method has some limitations and benefits.

5.1.1 Method and Materials

The platform utilized for these experiments was the 3DR Solo, as a part of work done with NYSEARCH. The 3DR Solo was chosen for several reasons, namely: (1) it has been proven to be a reliable platform with easy-to-use vetted software with little down-time for maintenance; (2) inexpensive spare parts; (3) lightweight (1.8 kg); and (4) uses an open-source Pixhawk autopilot that enables the research team to implement ancillary sensors like the ground-ranging LIDAR. Expanding further, the LIDAR sensor has been an invaluable measurement for getting accurate height information to better understand the relationship between the terrain data and the concentration and wind measurements. The Solo platform was modified to support an on-board in situ methane sensor placed in the joust configuration (see Section 2.1.2 for integration considerations). The methane sensor, a tunable diode laser absorption spectrometer (TDLAS), was entirely designed, built, and tested at NASA's Jet Propulsion Laboratory (JPL) – referred to as the open path laser spectrometer (OPLS). The OPLS' diode laser emits a mid-wave infrared (MWIR) light near 3.27 μm (3057 cm^{-1}). This particular band allows the OPLS to probe the strongest available absorption signals of methane (recall Fig. 1.4). Prior TDLAS instruments have utilized near-infrared (NIR) light bands at 1.6 μm for which the methane signal is nearly 50 $\times$ weaker as shown in Fig. 5.1.

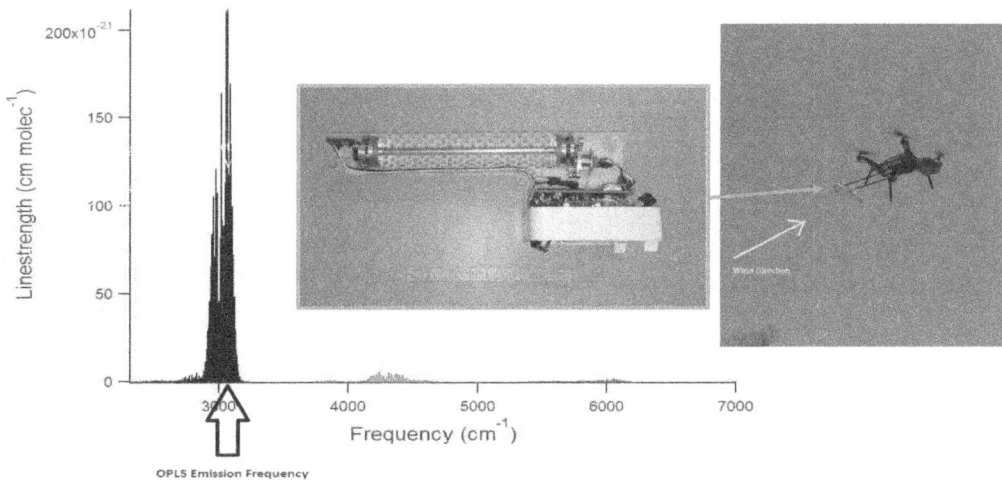

FIGURE 5.1 Image showing the OPLS emission frequency (3057 cm^{-1} = 3.27 μm), the OPLS sensor showing the open path optical head, and the integrated OPLS on the 3DR Solo quadcopter hovering and oriented into the wind.

To detect the emission source, the sUAS was flown downwind of the source. A general understanding of the Gaussian plume model (GPM) – see (4.3) – and atmospheric stability conditions can give guidance and/or intuition on where the plume should be and how wide. We know that the time-averaged plume of the emission source is oriented parallel to the mean wind direction. The stability class provides the conditions to promote the lateral and vertical dispersion. The temperature of the ground and air contribute to how the plume rises as it travels downwind. Combining these three understandings gives the operator a general sense of where to attempt plume detection. Once the detections are made, the source, in this case, is located in the reverse direction of the mean wind. The main idea is that a concentration packet (or group of molecules) travels in the direction of the wind, making a trajectory, to the sensing region. Assuming this path is approximately linear, the backward trajectory (or back-trajectory) method is applied to point a vector toward the source.

5.1.2 Rural Experiment

During February 14–15, 2017, a controlled release field experiment was conducted at the Merced Vernal Pools and Grassland Reserve (MVPGR). The MVPGR is located in Merced, CA, directly northeast of the UC Merced Campus (see Fig. 5.2). Temperatures during testing ranged from 13 °C to 18 °C. Wind speeds ranged from 0 to 3 m s^{-1}, generally from all directions. The combination of fairly low wind speeds and variable wind direction necessitated longer time in the field to adjust for this variability. There was no precipitation. Cloud coverage was 20% during the entire experiment. A cylinder of 100% methane was placed on the ground as shown in Fig. 5.2. A regulator was used to bring the pressure down to 15 PSI and the methane gas was flowed into a perforated Teflon tube embedded in coarse gravel as shown in Fig. 5.2 (top left).

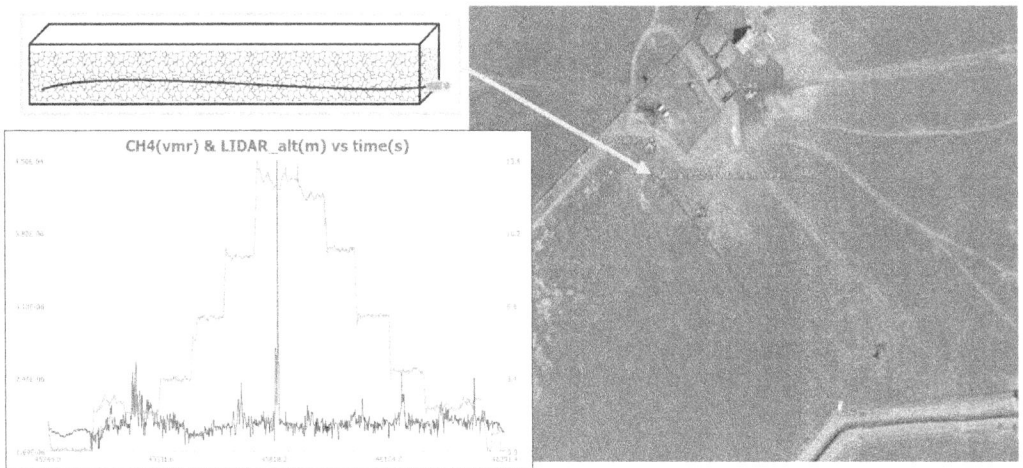

FIGURE 5.2 The right shows a satellite view of MVPGR (UC Merced) test site. The buildings to the north are an abandoned farm house. For reference, the line is 270 m long. On the left, shows the gas diffuser setup and data from a 30 m downwind flight with a 5 SCFH methane release.

TABLE 5.1 Testing matrix for February 14–15 Merced Vernal Pools controlled release test experiments. The flow rate is given in standard cubic feet per hour (SCFH).

Flt #	Date	Type	Description	Flow Rate	Mass Balance?
1	Feb 14	Downwind	SE of leak	5	Yes
2	Feb 14	Downwind	SE of leak	25	Yes
3	Feb 14	Downwind	SE of leak	25	Yes
4	Feb 14	Downwind	SE of leak	25	Yes
5	Feb 14	Downwind	NW of leak	25	Yes
6	Feb 14	Downwind	SE of leak	25	Yes
7	Feb 14	Downwind	SW of leak	5	Yes
8	Feb 14	Downwind	W of leak	5	Yes
9	Feb 14	Downwind	SW of leak	5	Yes
10	Feb 14	Downwind	NW of leak	5	Yes
11	Feb 14	Downwind	NW of leak	1	Yes
12	Feb 14	Downwind	NW of leak	1	Yes
13	Feb 14	Survey	near leak	1	No
14	Feb 15	Flt survey	near leak and cows	5	No
15	Feb 15	Flt survey	near leak	5	No
16	Feb 15	Downwind	NW of leak	0.2	No
17	Feb 15	Downwind	NW of leak	0.2	No
18	Feb 15	Flt survey	near leak	0.2	No
19	Feb 15	Downwind	NW & SW of leak	5	Yes
20	Feb 15	Flt survey	near leak	5	Yes
21	Feb 15	Flt survey	near leak	5	No
22	Feb 15	Flt survey	near leak	1	No
23	Feb 15	Flt survey	near leak	7	No
24	Feb 15	Flt survey	near leak	0.4	No
25	Feb 15	Flt survey	near leak	12	No
26	Feb 15	Flt survey	near farm house	NA	No
27	Feb 15	Flt survey	near farm house	NA	No

Flow rates ranged from 0.1 to 25 standard cubic feet per hour (SCFH). An ultrasonic anemometer (namely, R.M. Young 81000) was placed near the methane source. It transmitted wind data (direction and speed) wirelessly to the ground control station (GCS) where it was monitored by the GCS operator of the drone. This wind data was combined with the sUAS data (e.g. GPS and methane measurements) into a structured data set (i.e. interpolated to have the same timestamp and frequency – based on the sample frequency of the methane sensor).

The controlled release campaign consisted of 27 flights at several different leak rates (shown in Table. 5.1). Typically, flights were 30 meters away from the leak source at altitudes ranging from 3 to 25 meters above ground. Two types of flight profiles were flown. Downwind flights consisted of flying downwind of the leak source at 30 meters distance along a 50-m long line whose center point was parallel to the average wind direction for the preceding 5 min. For the downwind flights, several altitudes were flown in order to understand the vertical extent of the methane plume. For those cases where a discernible upper altitude could be determined, mass-balance analysis can be performed. Flight speeds were between 2 and 4 m s^{-1}. The sensor was always oriented facing the emission source. Average flight times were 9 minutes.

TABLE 5.2 Number of passes needed to detect a 5 SCFH leak with a 95% confidence given the Pasquill-Gifford atmospheric stability class above ground level (AGL) altitude, average wind speed, and distance to the source.

Date	Class	Wind (m/s)	Distance(m)	AGL(m)	Detection	95%
Feb. 2017	B	1.3	30	5	0.55	4
Feb. 2017	B	1.3	30	10	0.32	8
Aug. 2016	C	2.8	30	5	0.85	2
Aug. 2016	C	2.8	30	10	0.08	36

The flights conducted here occurred with wind speeds 1.3 ± 0.9 m s^{-1} with wind direction variance (σ_θ) of 65°. A typical flight pattern where different altitudes were investigated as shown in Fig. 5.2.

It is helpful to contrast the Feb. 14–15, 2017 measurements with measurements taken from an earlier experiment in August 2016 under similar conditions, in the same location, and with the same methane release rate. However, in the 2016 experiment, the wind conditions were more stable (2.8 ± 0.4 m s^{-1}, $\sigma_\theta = 25°$). In the 2016 case, the methane measurement signal above 10 m altitude was found to be nearly zero, whereas in the February 2017 experiment, 20% of the passes downwind of the emission resulted in a measurable methane signal (which can be observed in Fig. 5.2).

Data from both experiments were analyzed for the fraction of downwind passes that had clear leak indications. This meant that when flying 30 m downwind of the 5 SCFH leak, along a 50-m long line perpendicular to wind direction, we measured how often the methane signal > 100 ppb was observed (above background). During the laminar conditions of August 2016, at 5 m altitude, nearly 85% of passes indicated the presence of a leak. However, in the more turbulent, less atmospheric stable conditions of Feb. 2017, at 5 m altitude, only 55% of passes indicated a leak. Thus in order to obtain the number of passes required for a 95% confidence (i.e. the upwind area was appropriately sampled for leaks), the detection fractions can be related to the number of passes given the atmospheric conditions. This is seen for the two experiments described here in Table 5.2. An examination of these detection fractions as a function of altitude from both August 2016 and February 2017 are shown in Fig. 5.3. While unstable atmospheric conditions are not ideal to conduct downwind measurements, such conditions greatly increase the chance of detection at higher altitudes. Unstable atmospheric conditions also affected the effectiveness of leak localization. The challenges of atmospheric conditions are shown in Fig. 5.4 for 3 out of 15 downwind localization test flights. These three examples exemplify the range of results for localization. The perpendicular (shortest) distance from the leak source to the average back-trajectory line varied widely from 2 to 30 meters. In contrast, for the more laminar conditions of Aug. 2016, the average distances between the back-trajectory line and the leak location varied between 0 and 5 m.

To conclude, atmospheric conditions (i.e. wind speed and direction stability) contribute heavily in the process of detection. These conditions were observed to have a huge effect on the efficacy of the back-trajectory leak localization approach and on the due diligence leak detection for the experiments with average wind speed of 1.3 m s^{-1} versus 2.8 m s^{-1}.

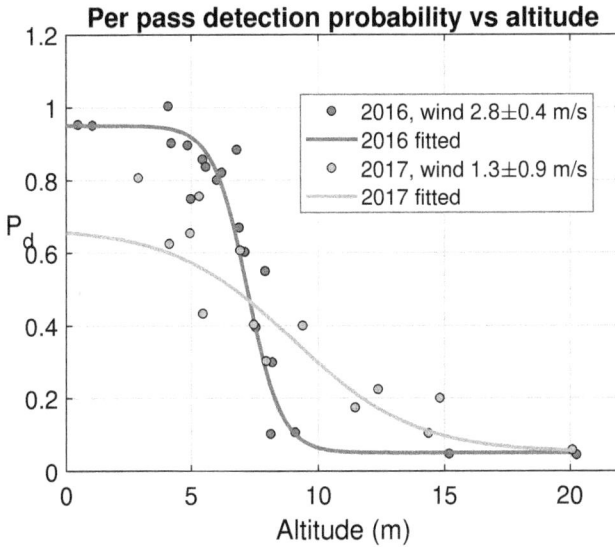

FIGURE 5.3 Contrast in measurements of two different wind conditions stable (2016) vs unstable (2017) under controlled methane release of 5 standard cubic feet per hour (SCFH) at 30 m downwind [4]. (reused from [3])

5.1.3 Urban Experiment

During March 27–29, 2017, flights were conducted at a simulated neighborhood at the Pacific Gas & Electric (PG&E) Livermore training facility in Livermore, California to test the detection and localization capability of the sUAS-based measurement system under different-sized continuous emission sources. The Livermore training

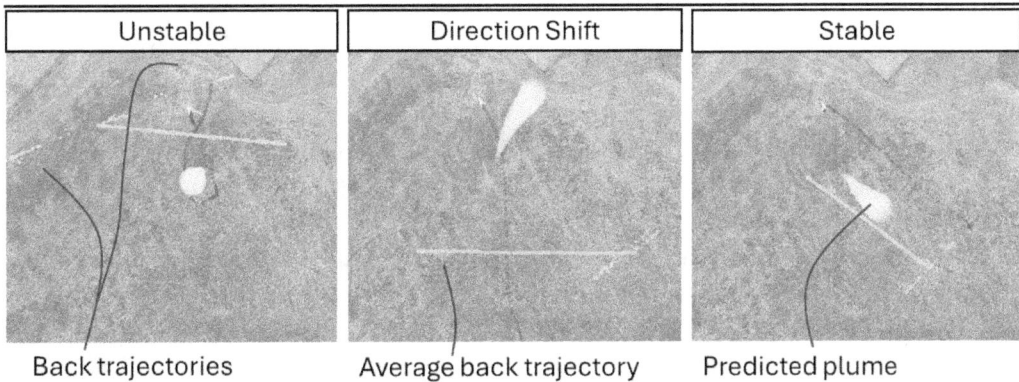

FIGURE 5.4 Four localization back-trajectory examples from the Feb. 2017 Merced field test are shown here. The short lines represent back-trajectories from individual methane signals and the bold long lines indicate the averaged back-trajectories for that flight. A snapshot of the predicted plume is shown emitting from where the leak location is, highlighting the temporal challenges with spatial-temporal measurements.

FIGURE 5.5   The PG&E Livermore training center. (left): The numbers refer to release points that map to the test matrix. (right): A map of the actual release points where all leaks were below-ground, except for L-1, L-4, L-9, L-18, and L-19 were above-ground.

facility allows for single and multiple controlled release points of natural gas from specific locations inside the simulated neighborhood (numbered 1–16 for readability of the test matrix). The neighborhood consist of 4 houses in each row and each house is located within 5 meters of the actual leak location (see Fig. 5.5). The experiment consisted of three consecutive test days, each with a slightly different testing matrix. On March 27 (day one), the leak locations of the test were shared between the test operators (NYSEARCH + PG&E) and the sUAS team (JPL and UC Merced). The first test consisted of simultaneously emitting one or two leaks at different locations in the neighborhood with a specified rate. On March 28 (day two), a series of blind tests were conducted, such that the test operators opened up one to four leaks and tasked the sUAS test team to identify the leak location(s) and number that were present. On March 29 (day three), one to four leak locations were shared between test operators and the sUAS team, simultaneously emitting during each test. The focus was on understanding the capabilities in large leak condition. The test matrix is shown in Table 5.3. Large leaks refer to leaks estimated to be > 10 SCFH. Medium leaks refer to leaks Pacific Gas & Electric 5–10 SCFH. Small leaks are < 5 SCFH. It should be noted that actual leak rates were not determined for these tests and leak rates were estimated from past experience by PG&E personnel.

The approach of the sUAS team for the tests was to conduct a sweeping survey across the whole 16 house area. Once completed, the areas with elevated methane signals were re-visited and investigated. During these tests, the issue of leak localization (identifying a lot or house) versus leak 'pinpointing' (defining a specific component or precise area) arose. On average, leaks were discovered and localized to ±2.5

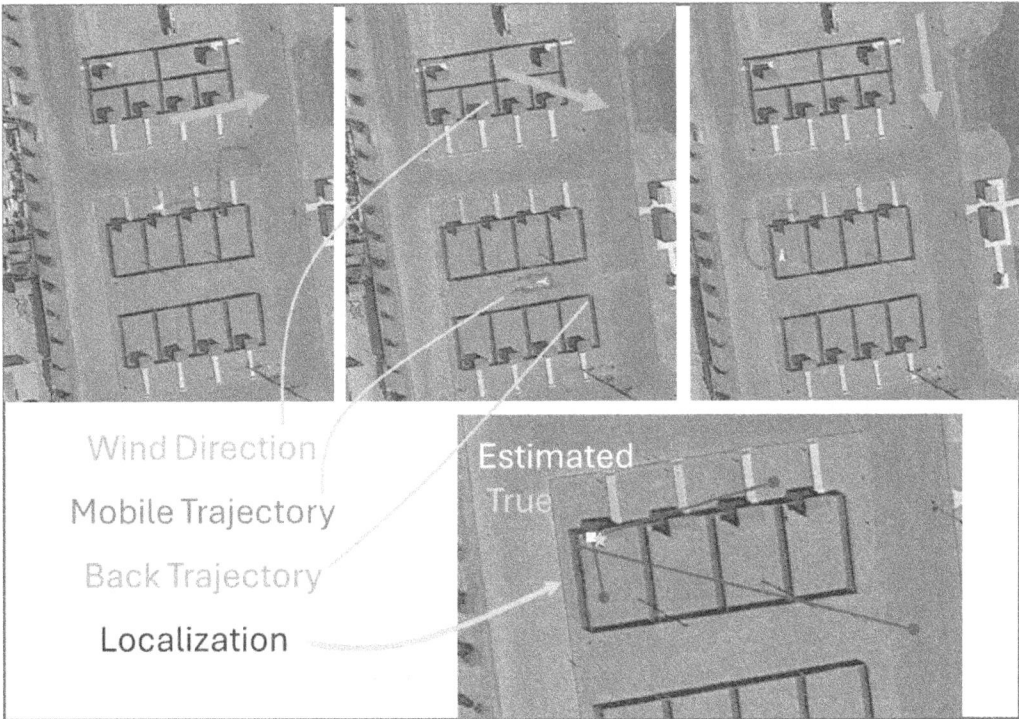

FIGURE 5.6 An example of the back trajectory method being heuristically applied in variable wind conditions, where the sUAS trajectory is given by small circles and the back trajectory vectors by the small lines. The bottom plot shows the averaged back trajectories (long lines) and the predicted leak location (square) versus the actual leak location (star) given: early flight back trajectories (top-left); mid-flight back trajectories (top-center); late back trajectories (top-right).

meters within 6 minutes, using the whole neighborhood scanning approach. However, to pinpoint the leak location to < 1 m, another 10+ minutes of flight time was typically required – often prompting a second flight to establish the desired accuracy. One of the major challenges with pinpointing the leak, was due to the rotorwash disturbing the ground. The added air disturbance often removed the presence of the methane signal or gave non-intuitive indications (e.g. contradictory methane signals compared to the initial survey) that were misleading. The sUAS team often had to back up (in the downwind direction) to re-establish the methane signal before beginning to trace the signal in the upwind direction to the source.

Other challenges observed during localizing the source, were with the stability in the wind speed and direction. There were several instances during tests where the wind direction shifted nearly $120°$ over the course of minutes (e.g. mid-flight). This turned out to be advantageous for leak detection because it enables the leak surveyor to triangulate the position of the leak. A good example of this can be seen in Test #5 (shown in Fig. 5.6). Early in the test, back trajectories of high methane signal suggested a leak location somewhere along a line near houses 5, 6, 7. Midway through the test, the wind shifted and the back trajectory indicated a leak location somewhere

TABLE 5.3   Test Matrix for PG&E Livermore Testing Facility Field Experiments.

Test #	Date	# Leaks	Leak Locations	Leak size	Blind
1	Mar 27	1	1	Large	No
2	Mar 27	1	9	Large	No
3	Mar 27	1	16	Large	No
4	Mar 27	1	4	Med	No
5	Mar 27	1	5	Med	No
6	Mar 27	1	13	Med	No
7	Mar 27	1	6	Small	No
8	Mar 27	1	7	Large	No
9	Mar 27	1	15	Medium	No
10	Mar 27	2	1, 8	Large	No
11	Mar 27	2	9,16	Large	No
12	Mar 27	2	4,8	Large	No
13	Mar 27	1	1	Med	No
14	Mar 28	1	6	Large	Yes
15	Mar 28	1	11	Med	Yes
16	Mar 28	2	4,16	Small	Yes
17	Mar 28	3	6,8,11	Small	Yes
18	Mar 28	3	5,10,13	Small	Yes
19	Mar 28	2	3,9	Small	Yes
20	Mar 28	4	4,6,10,15	Small	Yes
21	Mar 29	1	5	Large	No
22	Mar 29	4	5,6,7,8	Large	No
23	Mar 29	3	4,8,12	Large	No

along the line near houses 5 and 9. Where these lines intersect, was the west side of house 5. Later during the test, back trajectories confirmed the intersection point near the west side of house 5. Fig. 5.6 shows the average back trajectory vector for all these cases and compares it with the true leak location (yellow star). The predicted leak location is within 1 meter of the actual leak location.

During the seven blind tests that were conducted, the sUAS team used the same sweeping surveys across the whole area to identify areas of interest. Whenever a leak was encountered, the team used about 10–30 seconds to gather more methane signals to better determine where the leak may be coming from before moving on. Once the initial grid-based survey has been conducted, the areas of interest are re-investigated. Metrics of the blind surveys are tabulated in Table 5.4. The performance of our sUAS measurement system to detect and localize the blind test configurations, resulted in 14 of 16 leaks found. The leaks that were not found, were masked by other leaks in close proximity or found to be in-line with the wind direction.

5.2   CONTROLLED RELEASE EMISSION QUANTIFICATION

In this section we will highlight three controlled release experiments in two different environments, their results, and some insights learned in post-campaign fashion. The campaigns here include early work doing controlled release experiments at the MVPGR in 2017 (following the localization work) and in 2018 (focused more on quantification) as well as some more recent work at the Methane Emission Technology

TABLE 5.4 Metrics of blind test leaks. The time represents the time to localize all the leaks and the leak size is given in SCFH.

Test #	Leak Size	Time(min)	# leaks	# Leaks found
14	Large	16	1	1
15	Med	41	1	1
16	Small	28	2	2
17	Small	23	3	3
18	Small	24	3	2
19	Small	29	2	2
20	Small	31	4	3

Evaluation Center (METEC) in 2021. It is important to note that controlled release testing (such as those undertaken at METEC) provide a way to validate and verify the capabilities of sUAS-based systems along with other technologies. This gives practitioners and policy makes insight into how and when technologies can be used to provide compliance.

5.2.1 Method and Materials

There are three different sUAS platforms used in this section: a 3DR Solo, a DJI M210, and a DJI M300 (see Fig. 5.7). On all three platforms, a version of the OPLS (described in Section 5.1.1) was integrated. The only major difference between the OPLS versions was with the sampling rate. In later versions the OPLS, sampling rate was reduced to 5 Hz for increased performance. Aside from the Pixhawk system described earlier, the DJI sUAS platforms are integrated with the OPLS sensor through the downward facing skyport port using the onboard software development kit (SDK). The wind speed and direction in the first two cases were measured using the RM Young 3D ultrasonic anemometer, whereas the METEC test was implemented with two TriSonica ultrasonic anemometers. One of the TriSonica instruments

(a)

(b)

FIGURE 5.7 (a) DJI Matrice 210 (M210) equipped with OPLS developed by [1] and attached to the DJI Skyport. (b) DJI Matrice 300 (M300) equipped with the OPLS and TriSonica anemometer.

was installed at the OPLS sensor, and one was tripod mounted near the survey area of the METEC site.

5.2.2 February 2017 MVPGR

During the flights in the MVPGR (outlined in Section 5.1.2), the mass balance vertical flux plane method (see 4.1.3) was carried out to quantify a controlled point source emission released from ground level. The emission rate was varied from 1 to 25 SCFH. The total number of flights quantified consisted of 14, of which, some of the flights contained an ascending and descending flight path (counting as two individual estimations). The results can be seen in Fig. 5.9a, which shows that for most of the estimates, as turbulence intensity gets smaller, the estimates become better. This suggests that as the atmosphere becomes more stable the estimates become more accurate. From a theoretical perspective, this makes sense as the time averaged plume going through the measurement plane, would be governed by the mass balance equation. However, the atmospheric conditions tend to continuously change and provide disturbances to the estimates.

It is worth noting that the sequence of releases (shown in Table 5.1) goes from 5 SCFH all the way to 25 SCFH and eventually down to 1 SCFH. Therefore, it may be plausible that residual emissions from previous higher emission rates, resulted in over-estimations of the emission rate in the 1 SCFH case. In fact, from Fig. 5.9a it can be seen that if the 1 SCFH case was processed at a 5 SCFH emission rate, three of the estimates would be below 200% error – consistent with the rest of the estimates.

5.2.3 August 2018 MVPGR

In the same testing location as the February 2017 flights (in the MVPGR), the source setup included a pure methane bottle connected to a pressure regulator, followed by several meters of PFTE (Polytetrafluoroethylene) tubing, where a rotameter flowmeter gauge was placed for controlling the release rate. Additional PFTE tubing was used to connect the output of the flowmeter to a 5 Gallon bucket filled with coarse rocks ($\approx$ 1 cm in diameter). The rocks served to slow the stack velocity down and let the wind carry the methane, simulating a surface point source. The source rate was initially set to 10 SCFH, whereas after laboratory testing the source rate was calibrated to be $Q_* = 12.9$ SCFH. The landscape of the site was relatively flat with little topology and no surface obstructions (see Fig. 5.8). The atmospheric stability during the experiment was determined to be C and B (slightly to moderately unstable) using the PG curves [6]. The experiment consisted of 16 downwind flight paths of horizontal transects ascending from ground level till a clean air fetch was reached.

The results can be seen in Fig. 5.9b and in Table 7.1. It can be observed that as the turbulence intensity (TI) is decreasing, the accuracy of the mass balance method improves, which suggests that the atmospheric stability (more stable → smaller TI) is important for accuracy in the emission estimate as well as for improving the estimate variability. This result echos the result found in the previous section.

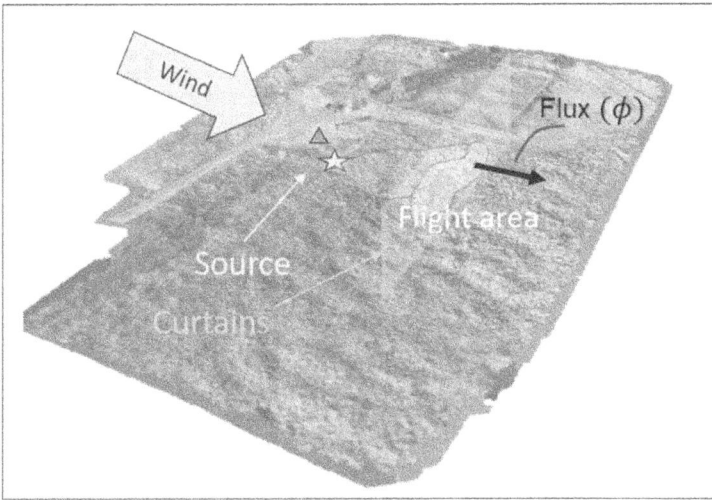

FIGURE 5.8   Experiment location at Merced Vernal Pools and Grassland Reserve. The star indicates the source, the triangle represents the location of the RMYoung anemometer, and the flight area is shown dotted line. (reused from [3])

5.2.4   April 2021 METEC

In April 2021, a series of controlled release experiments were undertaken with Aerometrix Services Inc. (Victoria BC, Canada) at the Methane Emission Technology Evaluation Center (METEC) over a five-day period. The controlled release tests were conducted concurrently with other teams on site simultaneously at each of the regions sub divided by the dirt roads within METEC. The testing configurations were changed every hour (called rounds) and the teams rotated to another sub-division or were given an hour rest. The 20 measurement rounds had mean measured methane

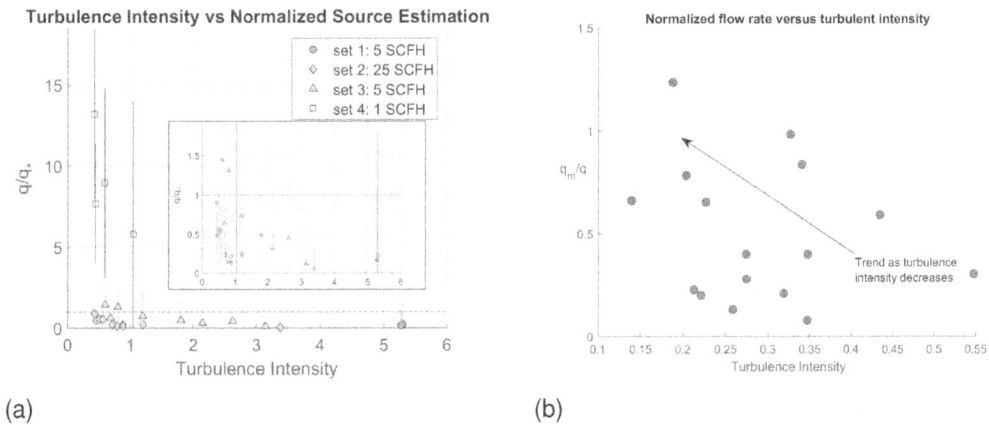

(a)   (b)

FIGURE 5.9   (a) Emission quantification from controlled release tests in the MVPGR February 2017. (b) Normalized controlled release estimation using the mass balance method of the August 2018 MVPGR field test.

concentrations ranging from 2.4 to 5.6 ppm – all significantly above the natural background. The emission detections at the drone, resulted in maximum methane concentrations from 3.1 to 152.8 ppm.

During this field campaign, we came across two notable challenges for measuring emissions in the open atmosphere with the OPLS-based sUAS approach. The first challenge has to do with off-site emissions, meaning, the emissions from adjacent sub-divisions of the METEC site that would meander into the area we are measuring. This is a challenge when the task is to detect, localize, and quantify within the sub-division itself. Therefore, the off-site emission needs to be tore out (or accounted for) of the emission estimate of the sub-divided site. This problem adds to the difficulty of measuring accurate emissions, due to the presence of higher emission rates, which leads to higher variability in the magnitude of the measurement signal. For example, given a large emission rate from an adjacent site and a relatively small emission from the target site, the variability between the difference in the upwind and downwind measurements (i.e. flux in and flux out of the control volume) could mask the presence of the smaller leak (see Fig. 5.10a). The other challenge we found with the

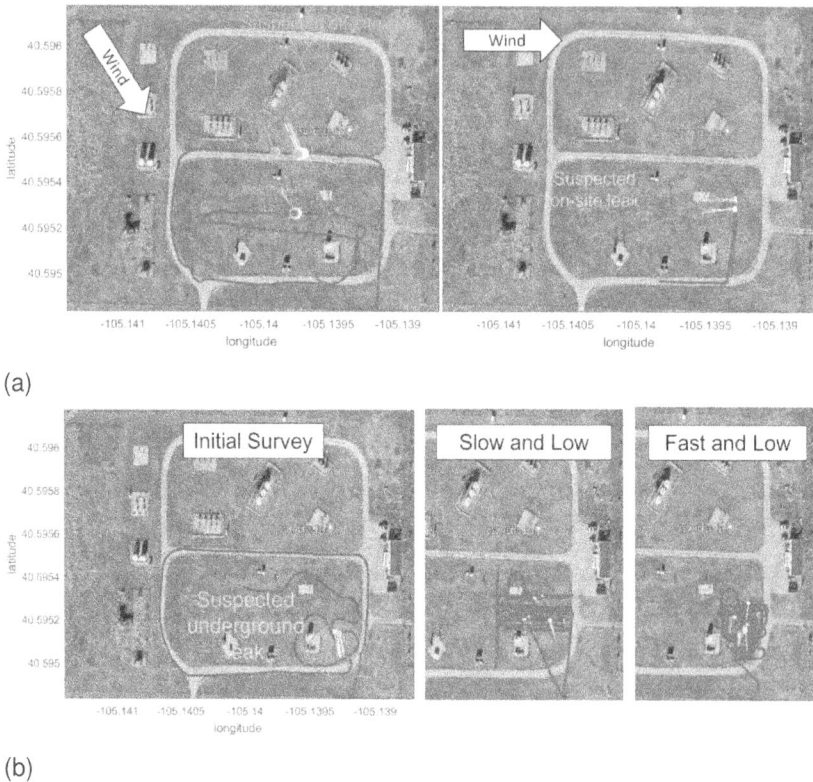

(a)

(b)

FIGURE 5.10 (a) Example of an off-site emission source affecting the survey and emission quantification. (b) Example of a survey with an underground source. The initial survey detected a leak near the south-east corner of the site. The successive flights had difficulty detecting the emission source when flying low and slow ($\approx$ 1m/s or less), whereas, when flying faster the rate of detection increases.

sUAS was with underground and surface-based leaks. For example, in Fig. 5.10b, the underground leak is first detected from the initial survey but when trying to quantify the leak, the detection became difficult when flying low to the ground with a slow flight speed ($< 1m/s$). One of the potential reasons for this could be due to the sUAS propeller wash interacting with the ground, causing the emission to be advected toward the ground and in an outward fashion from the aircraft (relative to the ground). Flying faster delays this effect, resulting in a better detection rate. This ultimately makes quantification difficult as the farther back we fly from the source, the more sensitive the measurement equipment needs to detect the emission. One advantage to measuring farther from the source is with the well-mixed condition. This provides a smooth concentration signal that is easier to estimate the source with (given the sensor is sensitive enough). Unfortunately, one of the constraints for the experiment was to only measure the source from inside the site fence line and sub divided region. The fence was done to mitigate the potential flight safety risk from sUAS and other personnel operating in an adjacent region.

5.3 REAL-WORLD EMISSION QUANTIFICATION

Once a quantification method has been developed, the overarching goals would be to subject them to real world applications. In this section we give examples of campaigns undertaken in real world environment scenarios.

5.3.1 Aug 2018: Merced County Regional Waste Management Landfill

The California Air Resources Board (CARB) introduced an implementation document to help municipal solid waste landfills be in compliance with California Code of Regulations title 17, subchapter 10, article 4, subarticle 6 sections 95460 to 95476, Methane Emissions from Municipal Solid Waste Landfills (MSWL). In short, this requires periodical regulatory surface emissions testing that may vary depending on the historical compliance of the site. Surface emissions monitoring (SEM) tests can take a considerable amount of time to complete (days to weeks) depending on the spatial scale of a site and how in-compliance the site has been previously. This is due to the fact that surveyors must traverse the site on foot. This requires the site to be broken into grids, which, can pose a resolution problem on how well the site is surveyed – potentially leading to emissions going undetected as well. Using alternative methods that rely on the plume dynamics of the surface emissions can allow for measurements to be taken adjacent to the site, such as with sUAS-based on-board measurements (as we have seen). Albeit the MSWL sites are much larger, the sUAS can be deployed using the vertical flux plane scanning approach and the emission rate be estimated using the mass balance approach (recall Chapter 4 for more details). Since sUAS cost is relatively small and can be deployed rapidly and more frequently than the traditional SEM surveys, we can see a potential to apply sUAS-based surveying to provide a faster alternative to leak detection as well as quantification.

In August of 2018, the Merced County Regional Waste Management (MCRWM) site was flown (see Fig. 5.11) with a DJI M210 equipped with the OPLS [1]. Due

FIGURE 5.11 At the MCRWM site a new cell is being developed, such that, the bank shown in the image has no membrane and only a dirt cover.

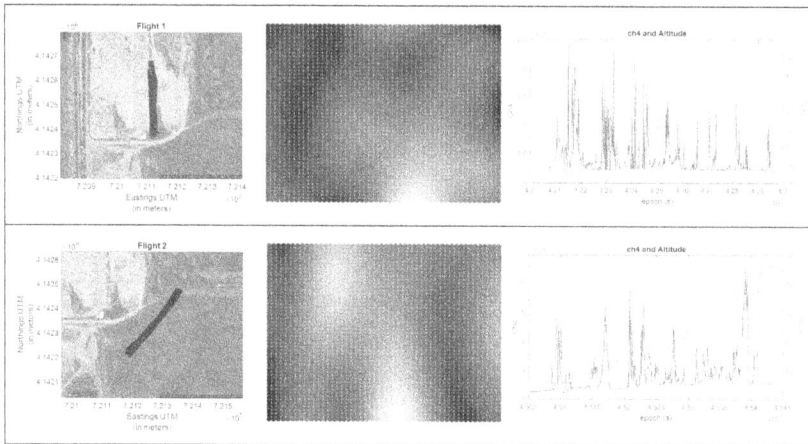

FIGURE 5.12 Two surveys of an active landfill cell without a membrane are conducted where measurements are taken from: (top) near the source; (bottom) further back. The measurements near the source show the emission emanating from the ground whereas when the measurement is taken from further downwind, the emission is subject to more turbulence.

to the limited amount of surveys conducted (only two flights) and the variability in the wind, the accuracy of the estimates yielded poor confidence, and the resulting emission estimates are likely off. However, the emission estimates for both flights were found to be above 200 SCFH (see Fig. 5.12), which is equivalent to the emissions of about 400 cows (given an average cow can typically produce 0.3–0.5 SCFH). One major challenge behind the use of sUAS, in this case, is that when measuring large

FIGURE 5.13   The mobile sUAS measurement platform above an eddy covariance tower in an Alaskan peatland bog. (reused from [5])

areas using the mass balance approach requires the atmosphere to be quite stable during the flight (which is on the order of the endurance of the measurement systems ≈ 20 minutes). Additionally, the size of the curtain makes the flight time requirement substantially larger than point source type measurements.

5.3.2   Nov 2019: Natural Emissions from Alaskan Permafrost Bog

Methane plays an important role in determining the atmosphere's climate and chemistry. Fluxes of methane from an ecosystem are often measured using eddy covariance flux towers (see Fig. 5.13), however, there are disadvantages with this method. Flux towers are expensive to purchase and have high demands with respect to maintenance and cost of operation, especially in remote locations, making replication across the landscape a challenge. In this case study, we applied the mass balance method to quantify the emissions from a permafrost bog using upwind and downwind quantification measurements [5].

During a field campaign in September of 2019, the Alaskan Peatland Experiment (APEX) site was conducted to explore the ability to measure natural ecosystem flux from the Alaskan peatland bogs. This work looked at using bootstrapping to understand the uncertainty better and experimented with different spatial interpolation methods (e.g. IDW, Kriging, see [2]) on the mass balance method. Additionally, this work explored the comparison between sUAS-derived wind (using an on-board anemometer) and the eddy covariance tower. An example of the flux plane measurements can be seen in Fig. 5.14 and the emission estimates can be seen in Fig. 5.15 and Table 5.5.

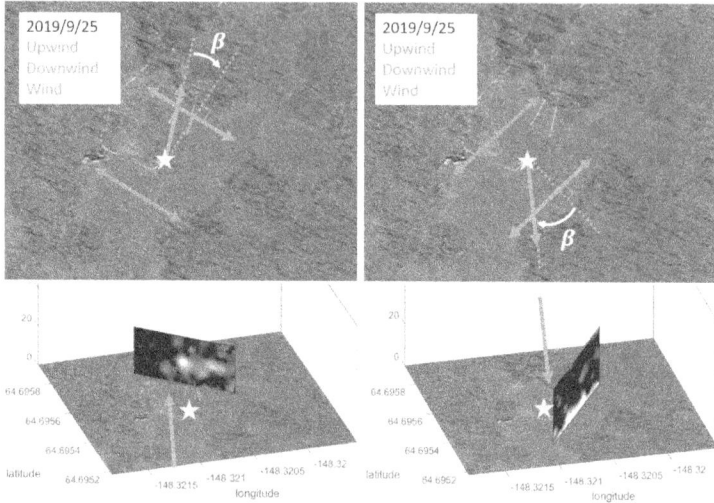

FIGURE 5.14 A top down view of the Alaskan Peatland Experiment (APEX) site, showing the upwind and downwind measurements for two different wind conditions. (reused from [5])

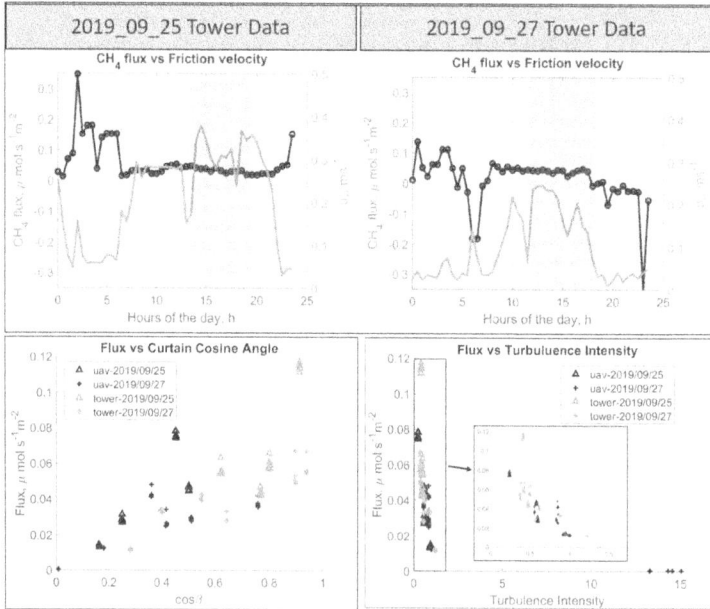

FIGURE 5.15 (top) Tower-based flux measurement of the APEX site for the days and times of the 2019 September field campaign. (bottom-left) Average flux given the cosine angle between the apparent wind direction and the normal vector of the flux plane. (bottom-right) The emission quantification values are shown to be higher when the turbulence intensity is small. (reused from [5])

TABLE 5.5 The quantification results are compared to tower estimates, which show that the sUAS-based estimates are lower than the tower measurements during the time of the flights. However, the estimates are within a reason considering the variability of the wind and the difficulty in measuring small emission rates.

Date	Avg Tower Flux	Avg sUAS Flux	sUAS Flux-1	sUAS Flux-2	sUAS Flux-3
9/25	0.0338 ± 0.0054	0.0220 ± 0.0278	0.0610 ± 0.0162	-0.0180 ± 0.0116	-
9/27	0.0411 ± 0.0063	0.0097 ± 0.0254	0.0078 ± 0.0094	0.0170 ± 0.0035	0.0045 ± 0.0126

Pause and Reflect

When it comes to real-world experimentation, it is always more challenging to control variables and achieve the results of our methodology given simulation. What are the advantages of being able detect, localize, or quantify emissions to the same degree as the simulation in real life?

5.4 CHAPTER SUMMARY

In this chapter, we have seen several experimental studies on detection, localization, and quantification of methane emissions using sUAS. In many of the examples, we observed that atmospheric conditions, primarily stability, play a key role in being able to accurately and precisely estimate the source location or quantify the emission source. We examined rural and urban localization experiments, rural and application-based controlled release quantification experiments, and real-world emissions from area sources and high and very low rates (i.e. natural emissions from permafrost). By exploring different localization and quantification methodologies and utilizing different sensing modalities, we can identify the best use cases for each application. For example, something that works for an oil and gas equipment pad, may not be the best setup for a landfill, and vice versa.

Bibliography

[1] Lance E Christensen. Miniature tunable laser spectrometer for detection of a trace gas, June 2017. US Patent 9,671,332.

[2] Derek Hollenbeck and YangQuan Chen. Characterization of ground-to-air emissions with sUAS using a digital twin framework. In *Proc. of the 2020 International Conference on Unmanned Aircraft Systems (ICUAS)*, pages 1162–1166. IEEE, 2020.

[3] Derek Hollenbeck and YangQuan Chen. A digital twin framework for environmental sensing with sUAS. *Journal of Intelligent & Robotic Systems*, 105(1):1, 2022.

[4] Derek Hollenbeck, Moataz Dahabra, Lance E Christensen, and YangQuan Chen. Data quality aware flight mission design for fugitive methane sniffing using fixed wing sUAS. In *Proc. of the 2019 International Conference on Unmanned Aircraft Systems (ICUAS)*, pages 813–818. IEEE, 2019.

[5] Derek Hollenbeck, Kristen Manies, YangQuan Chen, Dennis Baldocchi, Eugenie Euskirchen, and Lance Christensen. Evaluating a UAV-based mobile sensing system designed to quantify ecosystem-based methane. *Earth and Space Science Open Archive*, page 15, 2021.

[6] C Hunter. A recommended Pasquill-Gifford stability classification method for safety basis atmospheric dispersion modeling at SRS. Technical report, Savannah River Site (SRS), 2012.

II

Embedding Smartness to the Emission Source Determination Problem Solutions

CHAPTER 6

Digital Twin Framework

Embedding smartness to the emission source determination involves a broader use of modeling and data sharing with the physical system. In this chapter, we take a dive into the world of digital twins, building off of the Chapters 1–5 in part 1 of this book. We begin by giving a brief background on digital twins, including their definition and useful features. We will touch on how they are used to represent emission sources and their limitations. Then we will discuss the concept of behavior matching including the challenges associated with online-based optimization. Lastly, we will showcase the MOABS/DT platform for environmental monitoring applications.

6.1 AN INTRODUCTION TO DIGITAL TWINS

What is a Digital Twin? To begin we need to take a look at some of the early work using physical twins (or hardware twins). Physical twins can be seen as early as National Aeronautics and Space Administration's (NASA) Apollo program [2]. In this case, there existed at least two space vehicles or rovers at any given time. One of the rovers would be used in a controlled environment, and one of the rovers would be deployed in a real-world setting. This allowed for the continuous testing of code, maneuvers, and conditions before transmitting the executable code to the deployed version. This provided a sense of robustness and confidence in the control, that otherwise would not be present. Another great example is the 'iron bird'. The iron bird, which combined software and hardware together, created a physical interface of a plane cockpit with an aerodynamics simulation for training purposes. The simulation was used for a visual aid as well as to compute the meteorological conditions and resulting forces on the plane [2]. The concept of the early digital twins showed up in a product life-cycle management course at the University of Michigan in 2003 taught by Michael Grieves. It was represented as a digital equivalence of a physical product. Since then a white paper was written in reference to virtual factory replication [7]. It is explained that the DT contains three basic parts:

1. *physical* products in real space;

2. *virtual* products in virtual space; and

3. *connections* that tie together the real and virtual products.

DOI: 10.1201/9781003669470-6

Grieves contextualizes the use cases for DTs to be:

1. conceptualization – visualization of the physical and virtual products;

2. comparison – compare physical product behavior with the virtual product to improve results;

3. collaboration – solutions found can be applied immediately to other factories (or systems).

This vision for how DTs can be leveraged for improving performance of real systems or providing other insights was developed further by the community for smart cities [6], smart control engineering [28], etc. In general, a DT model, given the current implementation, can be classified into one of the four key levels of development:

(L1) Pre-Digital Twin;

(L2) Digital Twin;

(L3) Adaptive Digital Twin;

(L4) Intelligent Digital Twin.

Each level has increasing amounts of model sophistication and complexity. Aside from L1, all of the levels have a physical twin. As the DT levels go from L2 to L4 the data acquisition from the physical twin increases. Different system health and performance parameters can be extracted for a predictive maintenance standpoint (from batch updates to real-time updates). Additionally, these stages can start to incorporate machine learning (ML) in operator preferences as well as for the system/environment [18]. See Table 6.1 for a summary of the DT Levels.

6.2 REPRESENTING THE DIGITAL TWIN AS AN EMISSION SOURCE

In the literature, there are several ways we can approach the representation of an emission source, such as through first principle, stochastic models, hybrid models, and parameterized models. In this section we will highlight some of the key differences and limitations to each approach.

6.2.1 First principles

The first way, arguably the most accurate and expensive, is through first principles. This means to derive a set of partial differential equations (PDE) from general laws such as the continuity equation, momentum, or the energy balance. The goal is to solve the PDE that governs the particular system. For example, the advection and diffusion equation (ADE) with a source term (recall equation (3.1) from Chapter 3) can be used to describe a general emission source. The wind field can be described by the Navier Stokes equation – or by various other versions (e.g. Reynolds averaged NS, or Large Eddy Simulations). The wind field (or the advection field) can be used to propagate the ADE in time, simulating the source. The drawback is that these

TABLE 6.1 The generalized digital twin levels and associated capabilities and features.

Level	System Modeling	Physical System	System Interaction	Smart Capabilities	Features
L1	DT Environment based on general model	Not built	Not built	N/A	Preliminary Design
L2	DT Environment based on physical system	Operating standalone	No real-time data acquisition	No	Preliminary Design Performance analysis and system status
L3	DT Environment based on physical system with monitoring interface and parameter estimation	operating standalone with supervised systems	Real-time data acquisition	Limited	Preliminary Design Performance analysis and system status Data analytics, fault detection, prognosis, etc.
L4	DT Environment based on physical system with monitoring interface and adaptive behavioral learning	Operating in the loop with the DT virtual environment	Real-time data acquisition and smart control	Total	Preliminary Design Performance analysis and system status Data analytics, fault detection, prognosis, etc. Automated recommendations, and actions over the physical system

models require a huge amount of resources to run at high fidelity. The PDE is often broken down into a high-dimensional ordinary differential equation (ODE) problem and solved implicitly or explicitly. The computational domain, pre-defined boundary conditions, and implicit or explicit scheme have to be determined prior to the initial start.

In general, there are numerical limitations to solving PDEs, depending on which scheme is chosen. For example, given the viscous Burgers equation,

$$\frac{\partial u}{\partial t} = -u\frac{\partial u}{\partial x_i} + K\frac{\partial^2 u}{\partial x_i^2}, \tag{6.1}$$

an alternating direction implicit method can be used to achieve an unconditional convergence of the solution [20]. This means that the time step (Δt) can be chosen as large as you want. Alternatively, if we choose an explicit scheme (i.e. using a central difference stencil), the convergence is subject to the Courant–Friedrichs–Lewy (CFL) number (e.g. $u\Delta t/\Delta x$ for the linear convection problem). This puts constraints on the grid size and allowable time step size, rendering the computation slow. Furthermore, it can often lead to numerical instabilities. To illustrate, let's take a look at the differences between explicit and implicit formulations. For the ADE, a common explicit discretization uses the Forward Euler method in time and a central difference scheme for spatial derivatives,

$$\frac{u_i^{n+1} - u_i^n}{\Delta t} + v\frac{u_{i+1}^n - u_{i-1}^n}{2\Delta x} = D\frac{u_{i+1}^n - 2u_i^n + u_{i-1}^n}{\Delta x^2}. \tag{6.2}$$

Rearranging the equation such that all the $n+1$ terms are on the left hand side and the n terms on the right,

$$u_i^{n+1} = u_i^n - \frac{v\Delta t}{2\Delta x}(u_{i+1}^n - u_{i-1}^n) + \frac{D\Delta t}{\Delta x^2}(u_{i+1}^n - 2u_i^n + u_{i-1}^n). \tag{6.3}$$

Then for each i we can form a matrix A_{exp} such that, $\mathbf{u}^{n+1} = A_{exp}\mathbf{u}^n$. Whereas, a fully implicit scheme uses Backward Euler in time,

$$\frac{u_i^{n+1} - u_i^n}{\Delta t} + v\frac{u_{i+1}^{n+1} - u_{i-1}^{n+1}}{2\Delta x} = D\frac{u_{i+1}^{n+1} - 2u_i^{n+1} + u_{i-1}^{n+1}}{\Delta x^2}. \tag{6.4}$$

Rearranging the equation such that all the $n+1$ terms are on the left hand side and the n terms on the right,

$$u_i^{n+1} - \frac{v\Delta t}{2\Delta x}(u_{i+1}^{n+1} - u_{i-1}^{n+1}) + \frac{D\Delta t}{\Delta x^2}(u_{i+1}^{n+1} - 2u_i^{n+1} + u_{i-1}^{n+1}) = u_i^n, \tag{6.5}$$

we can rearrange into matrix form, such that, $A_{imp}\mathbf{u}^{n+1} = \mathbf{u}^n$. The differences between the two can be seen in the stability of the scheme (generally computed with the von Neumann stability analysis). The explicit scheme is found to be conditionally stable, requiring:

$$\frac{v\Delta t}{\Delta x} \le 1 \quad \text{(CFL condition)}, \tag{6.6}$$

TABLE 6.2 Performance of numerical schemes in the advection diffusion scenario.

Method	Time Accuracy	Space Accuracy	Stability
Explicit (Forward Euler)	$\mathcal{O}(\Delta t)$	$\mathcal{O}(\Delta x^2)$	Conditionally stable
Implicit (Backward Euler)	$\mathcal{O}(\Delta t)$	$\mathcal{O}(\Delta x^2)$	Unconditionally stable
Crank-Nicolson	$\mathcal{O}(\Delta t^2)$	$\mathcal{O}(\Delta x^2)$	Unconditionally stable but oscillatory

$$\frac{D\Delta t}{\Delta x^2} \leq \frac{1}{2} \quad \text{(diffusive stability).} \tag{6.7}$$

The implicit scheme however, is unconditionally stable (meaning we have no restrictions on the size of Δx and Δt). The precision of these methods can be seen in Table 6.2. The interested reader should consult [23].

Other notable ways to compute the solution of the PDE is through the use of spectral methods. Spectral methods are a class of numerical techniques used to solve partial differential equations (PDEs) by expanding the solution in terms of global basis functions (such as Fourier series or Chebyshev polynomials). These methods are known for their high accuracy when solving problems with smooth solutions due to their exponential convergence properties. They begin by approximating the solution as a sum of basis functions,

$$u(x,t) \approx \sum_{K=0}^{N} a_k \phi_k(x), \tag{6.8}$$

where a_k represents the spectral coefficients and $\phi_k(x)$ represents the basis functions. Given periodic boundary conditions, $\phi_k(x)$ can be represented by the Fourier basis functions,

$$\phi_k(x) = \exp(ikx). \tag{6.9}$$

If the PDE has a finite domain with boundary conditions, Chebyshev polynomials can be utilized,

$$\phi_k(x) = T_k(x). \tag{6.10}$$

Once the basis is chosen, it is substituted into the PDE directly or by Galerkin projection. Then the PDE becomes a system of ODEs that can be solved using Runge-Kutta or Crank-Nicolson methods for time-stepping the solution forward.

To illustrate further, let's suppose we use a Fourier basis in solving (6.1) with no advection term. Applying (6.8) to both sides, the left and right side can be computed as,

$$\frac{\partial^2 u}{\partial x^2} = \sum a_k(t)(ik)^2 \exp(ikx), \tag{6.11}$$

$$\frac{\partial u}{\partial t} = \sum \frac{\partial a_k(t)}{\partial t} \exp(ikx). \tag{6.12}$$

Equating coefficients we find,

$$\frac{\partial a_k(t)}{\partial t} = -Da_k(t)k^2, \tag{6.13}$$

where the coefficients $a_k(t)$ are computed as,

$$a_k(t) = \int_\Omega u(x,t)\exp(-ikx)dx \tag{6.14}$$

from the initial condition $u(x,0) = f(x)$, the initial coefficient is,

$$a_k(0) = \int_\Omega f(x)\exp(-ikx). \tag{6.15}$$

It is important to know that in this case the boundaries are periodic, $u(L,t) = u(-L,t)$, for example, if $-L \le x \le L$.

6.2.2 Stochastic models

The second way for solving the emission source is by Lagrangian formulation and stochastic integration of a stochastic differential equation. For example, the Langevin equation can be used to sequentially propagate a particle subject to a random noise and drift. The drift in this case is typically described by the wind field and the noise is chosen to be white noise (or Weiner process). Take for example, the diffusion equation for the concentration field $y(\mathbf{x},t)$ in the absence of advection is given by,

$$\frac{\partial y}{\partial t} = D\nabla^2 y + Q\delta(\mathbf{x}), \tag{6.16}$$

where D is the diffusion coefficient. The Green's function solution to this equation in 2D is given as,

$$y = \frac{Q}{4\pi Dt}\exp\left(-\frac{x_1^2 + x_2^2}{4Dt}\right), \tag{6.17}$$

with Q being the source rate. Switching to polar coordinates, where $r^2 = x_1^2 + x_2^2$, we can write the mean square displacement (MSD) as

$$\langle r^2 \rangle = \int_{-\infty}^{\infty}\int_0^{2\pi} r^2 \frac{Q}{4\pi Dt}\exp\left(-\frac{r^2}{4Dt}\right)rd\theta dr. \tag{6.18}$$

It follows that the solution to this integral is, $\langle r^2 \rangle = 4Dt$. This process can be modeled stochastically using the Langevin equation, such that,

$$\frac{d\mathbf{x}}{dt} = \xi(t), \tag{6.19}$$

where $\xi(t)$ is a Gaussian white noise process with properties, $\langle \xi_i(t) \rangle = 0$, and

$$\langle \xi_i(t)\xi_j(t') \rangle = 2D\delta_{ij}\delta(t - t'). \tag{6.20}$$

The MSD of a diffusing particle is then computed as,

$$\langle |\mathbf{x}(t) - \mathbf{x}_0|^2 \rangle = 2dDt, \tag{6.21}$$

where d is the number of spatial dimensions. Comparing to the 2D analytical solution, we see that the MSD behaves the same. Now, what if the system is in the presence of an advecting velocity field $\mathbf{u}(\mathbf{x}, t)$? The governing equation can then be written as

$$\frac{\partial y}{\partial t} + \mathbf{u} \cdot \nabla y = D\nabla^2 y. \tag{6.22}$$

The corresponding Langevin equation with a deterministic drift term is:

$$\frac{d\mathbf{x}}{dt} = \mathbf{u}(\mathbf{x}, t) + \xi(t). \tag{6.23}$$

Here, $\mathbf{u}(\mathbf{x}, t)$ represents the bulk transport of particles, while $\xi(t)$ models small-scale random motion due to diffusion. To analyze the spread of particles in an advective flow, we compute the MSD

$$\langle |\mathbf{x}(t) - \mathbf{x}_0|^2 \rangle = |\mathbf{u}t|^2 + 2dDt. \tag{6.24}$$

This equation shows that in an advective flow, the spread of particles consists of a deterministic term $|\mathbf{u}t|^2$ due to advection and a diffusive term $2dDt$.

6.2.3 Parameterized models

Given specific scenarios, such as time-averaged concentrations, the emission source can be represented by a parameterized model. A popular case is the Gaussian plume model (or GPM – recall equation (4.3) in Chapter 4). The model can be solved by utilizing the Laplace transform,

$$\mathcal{L}\{f(t)\} = F(s) = \int_{0^+}^{\infty} f(t)e^{-st}dt, \tag{6.25}$$

subject to some assumptions, and the proper initial and boundary conditions. Assumptions such as: neglecting the diffusion in the x_1 direction, steady state $\frac{\partial y}{\partial t} \to 0$, and constant diffusion coefficient $\mathbf{D}(\mathbf{x}) = [D_1(x_1), D_2(x_2), D_3(x_3)]^T \to D$. The equation can be simplified to

$$u_1 \frac{\partial y}{\partial x_1} = D\left(\frac{\partial^2 y}{\partial x_2^2} + \frac{\partial^2 y}{\partial x_3^2}\right). \tag{6.26}$$

Following the derivation outlined in [26], the variables can all be normalized by: $\tilde{x}_1 = (D/uH^2)x_1$, $\tilde{x}_2 = x_2/H$, $\tilde{x}_3 = x_3/H$, and $\tilde{y} = (uH^2/Q)y$. The variable x_1 can be replaced by

$$r = \frac{1}{u} \int_0^{x_1} D(\xi)d\xi, \tag{6.27}$$

which further simplifies the equation by

$$\frac{\partial y}{\partial r} = \frac{\partial^2 y}{\partial x_2^2} + \frac{\partial^2 y}{\partial x_3^2}. \tag{6.28}$$

These models traditionally look at emissions from an effective stack height (height plus plume rise), $H = h + \delta h$, and associate the stack height as a design parameter when evaluating the air quality impacts nearby. Using a separation of variables approach,

$$y(r, x_2, x_3) = \frac{Q}{u} D_2(r, x_2) D_3(r, x_3). \tag{6.29}$$

This leads to two reduced dimensional problems with the following initial and boundary conditions,

$$\frac{\partial D_2}{\partial r} = \frac{\partial^2 D_2}{\partial x_2^2}, \quad \text{for } 0 \leq r < \infty, \text{ and } -\infty < x_2 < \infty, \tag{6.30}$$

$$D_2(0, x_2) = \delta(x_2), \quad D_2(\infty, x_2) = 0, \quad D_2(r, \pm\infty) = 0, \tag{6.31}$$

and

$$\frac{\partial D_3}{\partial r} = \frac{\partial^2 D_3}{\partial x_3^2}, \quad \text{for } 0 \leq r < \infty, \text{ and } 0 \leq x_3 < \infty, \tag{6.32}$$

$$D_3(0, x_3) = \delta(x_3 - H), \quad D_3(\infty, x_3) = 0, \quad D_3(r, \infty) = 0, \tag{6.33}$$

$$\frac{\partial D_3}{\partial x_3}(r, 0) = 0. \tag{6.34}$$

The derivation can be seen in [26], however, the solution is given as

$$y(r, x_2, x_3) = \frac{Q}{4\pi u r} \exp\left(-\frac{x_2^2}{4r}\right) \left[\exp\left(-\frac{(x_3 - H)^2}{4r}\right) + \exp\left(-\frac{(x_3 + H)^2}{4r}\right)\right]. \tag{6.35}$$

This is different to the traditional GPM model shown in (4.3). This is primarily because the standard deviation of the plume is defined by

$$\sigma^2(x_1) = \frac{2}{u} \int_0^{x_1} D(\xi) d\xi = 2r, \tag{6.36}$$

which replaces x_1 in equation (6.26) for instance. There are several functional forms proposed in the literature, including a simple power law expression, $\sigma^2(x_1) = ax_1^b$, where the coefficients are determined experimentally or through a lookup table.

6.2.4 Data-driven and Machine Learning models

Scientific Machine Learning (SciML) [27] and Physics-Informed Machine Learning (PIML) [13] have emerged as powerful approaches for solving partial differential equations (PDEs) and modeling complex physical systems by integrating physics-based constraints with data-driven methods. Unlike purely numerical solvers, these approaches leverage machine learning architectures to approximate solutions efficiently while ensuring physical consistency. This is particularly useful for real-time simulations in digital twin frameworks, where models must update dynamically based on new sensor measurements. A key advantage of PIML is its ability to incorporate prior knowledge of governing equations directly into the learning process, reducing the need for large labeled datasets and improving generalization to different problem settings

[13]. Recent advances in SciML have emphasized the importance of training set diversity when approximating PDE solutions across different parameter regimes. Studies show that training ML models on a broad range of initial conditions, boundary conditions, and PDE coefficients improves interpolation and extrapolation. Additionally, curriculum learning – where models first learn simple PDE cases before progressing to complex ones – has been shown to enhance convergence and accuracy [14, 27].

Among the various deep learning architectures used in SciML, Fourier Neural Operators (FNOs) are particularly effective for learning solution mappings in infinite-dimensional function spaces. FNOs employ the Fast Fourier Transform (FFT) to learn spectral representations of PDE solutions, making them mesh-independent and well-suited for problems with variable coefficients or complex geometries [15]. This approach has shown promising results in applications such as fluid dynamics and weather modeling, where traditional solvers can be computationally expensive. Similarly, DeepONets (Deep Operator Networks) decompose PDE solutions into a trunk network, which encodes the spatial domain and a branch network which encodes problem parameters, allowing efficient representation of PDE solution operators [17]. Since DeepONets operate on function spaces rather than discrete grids, they are particularly useful for real-time inference in digital twins, enabling fast predictions for systems with continuously varying parameters.

In contrast to these operator-learning methods, Physics-Informed Neural Networks (PINN) directly encode the governing PDEs into the loss function

$$\mathcal{L} = \mathcal{L}_{data} + \lambda \mathcal{L}_{PDE}, \tag{6.37}$$

ensuring both equation residuals and data consistency are minimized [24]. By incorporating boundary and initial conditions within the learning process, PINNs can infer solutions to PDEs even in cases where sparse or noisy data are available. However, PINNs are known to suffer from training inefficiencies, particularly for highly stiff equations or multiscale problems, requiring advanced training strategies such as adaptive loss weighting or domain decomposition [29].

In digital twin applications, these machine learning models enable real-time simulation, uncertainty quantification, and adaptive learning. Traditional PDE solvers often struggle with computational costs, making them impractical for continuous system monitoring. FNOs and DeepONets provide rapid PDE evaluations, allowing digital twins to update their state estimates dynamically. Moreover, uncertainty quantification techniques, such as Physics-Informed Gaussian Processes (PIGPs) and Bayesian PINNs, allow digital twins to assess confidence in their predictions, which is crucial in environmental monitoring and industrial process control [30]. Another critical advantage of these approaches is their ability to tackle inverse problems, where unknown system parameters – such as emission source locations in methane detection—are inferred from observational data. In such applications, PINNs and DeepONets can estimate system parameters with improved accuracy compared to traditional optimization-based techniques [3].

Despite these advantages, SciML and PIML face several challenges. PINNs often struggle with slow convergence and vanishing gradients, particularly when solving PDEs with high-frequency components. FNOs and DeepONets, while efficient in their function-space representations, require careful selection of training data to generalize effectively across different PDE parameter regimes. Recent studies suggest that diverse training sets, covering a broad range of boundary conditions and physical properties, are essential for robust model performance [1]. Additionally, hybrid approaches that combine traditional solvers with ML acceleration – such as physics-informed recurrent neural networks (PI-RNN) or coarse-grid solvers enhanced by deep learning show promise in overcoming some of these limitations [16].

Overall, SciML and PIML provide a transformative framework for solving PDEs, bridging the gap between physics-based numerical models and modern deep learning techniques. As research advances, the integration of meta-learning, active learning, and hybrid numerical-ML solvers will further enhance the reliability of these approaches for real-world applications. In digital twins, these methods will play a crucial role in enabling real-time predictions, optimizing control strategies, and improving decision-making in complex engineering and scientific systems.

6.2.5 Hybrid models

Hybrid modeling – Hybrid modeling is an approach that combines physics-based models, numerical simulations, stochastic methods, and machine learning techniques to improve accuracy, generalization ability, and computational efficiency in solving complex systems governed by partial differential equations or other mathematical frameworks. The advantages and disadvantages of modeling using deterministic (first principles – Eulerian point of view) or through stochastic (Lagrangian point of view), hybrid models can be utilized and avoided based on the way the overall system is created. We will see an example in Section 6.4.

6.3 BEHAVIOR MATCHING

Once a Digital Twin model is chosen and a physical system exists, we collect output data and perform behavior matching. **Behavior matching** – is the process by which the output from the physical system and the output from the digital twin are compared iteratively, until the digital twin output sufficiently represents the output of the physical system (see Fig. 6.1). This means that during the comparison between the observed data and the modeled data, the digital twin parameters are tuned to match its behavior to that of the physical system. The behavior matching process can be typically viewed as an inverse problem and is highly dependent on how the digital twin model is implemented. As shown in the previous section, the digital twin forward model can be expressed as a numerical, stochastic, analytic/parametric, machine learning, or hybrid model/system. Similar to optimization in general, there are two ways to perform behavior matching: online and offline. Online and offline behavior matching represent two distinct approaches to parameter estimation in inverse problems, each with unique challenges and opportunities. In this section we will discuss them in general while in Chapter 8 we will discuss inverse problems and ill-posedness.

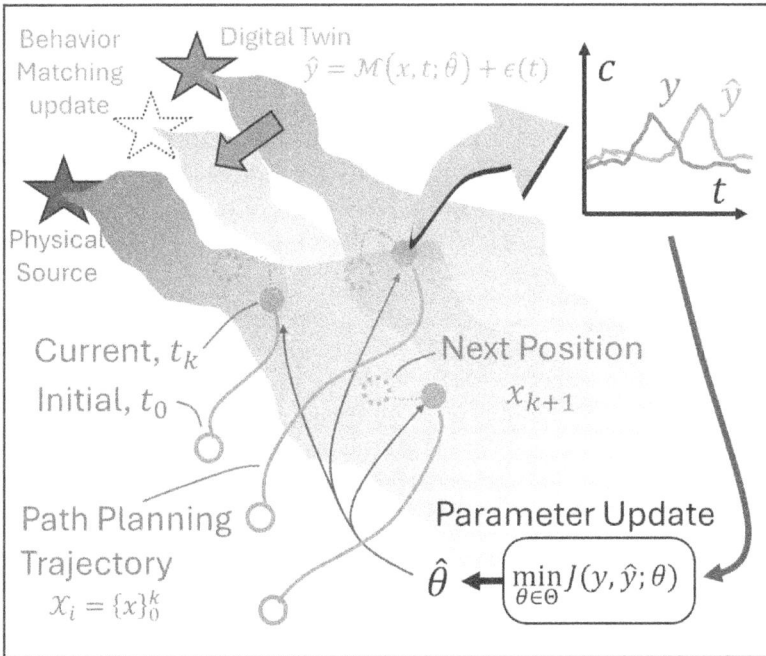

FIGURE 6.1 An example of behavior matching problem given a trajectory of mobile sensor data.

6.3.1 Offline Behavior Matching

Offline Behavior Matching – Offline behavior matching involves parameter estimation using pre-collected data, typically through batch processing. If time is not as sensitive for the digital twin application, using an offline-sense behavior matching would allow for data to be collected over a long period, or under different use cases. This approach may work well for the initial parameter tuning of the digital twin. Once tuned, the digital twin can be deployed for tasks such as performance assessment or prognosis. The main advantage of this approach is that it allows for computationally intensive methods, such as Bayesian inference, variational optimization, or deep learning-based surrogate modeling, to be employed without real-time constraints. Given a model $\mathcal{M}(\mathbf{x}, \theta)$ parameterized by θ, the objective is to find the optimal parameters that minimize a discrepancy measure between model outputs and observed data. A general approach is to construct a cost function $J(\cdot)$ to optimize over. Many practitioners utilize the least squares approach by letting the cost function be,

$$\theta^* = \arg\min_{\theta \in \Theta} J(\mathbf{y}, \hat{\mathbf{y}}, \mathbf{x}, \theta) = \arg\min_{\theta \in \Theta} \sum_{i=1}^{N} ||y_i - \mathcal{M}(\mathbf{x}_i, \theta)||^2, \qquad (6.38)$$

where y_i are the i-th observed measurements and $\mathbf{x}_i$ represents the i-th spatial and/or temporal locations. The modeled data is given by $\hat{\mathbf{y}} = \mathcal{M}(\mathbf{x}, \theta) + \eta(t)$. Offline approaches can leverage extensive training data and advanced optimization techniques such as genetic algorithms, simplex search (or Nelder-Mead),

Broyden–Fletcher–Goldfarb–Shanno (BFGS), adjoint methods [19], or gradient-based optimization (e.g. gradient descent, stochastic gradient descent, accelerated gradient descent, triple momentum, etc.) [21] via automatic differentiation. A key challenge is ensuring that the inferred parameters generalize well to new conditions. This issue has been highlighted in scientific machine learning (SciML), where the diversity of training data plays a crucial role in model robustness and generalization. Additionally, offline methods struggle when the underlying system is highly dynamic, as they cannot adapt to changes in real-time.

6.3.2 Online Behavior Matching

On the other hand, the online-sense approach requires the measurement data to be streamed to the location where the digital twin is being stored and updated. This may require additional steps to retrieve the observation. Then, depending on the physical system of interest, the chosen model needs to be used to predict or estimate the new parameters in an efficient and accurate way. The computational cost depends heavily on the complexity and computational efficiency of the digital twin model. When the digital twin becomes very high order and is not in closed form, the process of optimizing model parameters becomes increasingly difficult and sometimes intractable in real-time. Therefore, having models with fast computation are critical in this type of behavior matching problem.

Online Behavior Matching – In contrast, online behavior matching estimates parameters dynamically as new measurements become available. This is typically done via recursive estimation techniques, such as Kalman filtering, ensemble data assimilation, or real-time optimization. The general form of the online estimation problem involves updating parameters at each time step $t = \Delta tk$ using new observations y_k,

$$\theta_{k+1} = \theta_k + K_k(y_k - \mathcal{M}(\mathbf{x}_k, \theta_k)), \tag{6.39}$$

where K_t is a gain matrix (as in the Kalman filter) or an adaptive learning rate in machine learning-based implementations. One major advantage of online behavior matching is its ability to adapt to changing system dynamics, making it particularly useful for real-time monitoring and control applications, such as methane emission detection via drone-based sensing. However, online approaches must balance computational efficiency with accuracy, as updates need to be performed within stringent time constraints. Furthermore, online methods are often sensitive to measurement noise, requiring robust filtering techniques to ensure stable convergence.

Comparative Challenges and Opportunities – A fundamental trade-off exists between computational cost and adaptability. Offline methods can afford to use complex numerical simulations or machine learning models trained on large datasets, whereas online methods require simplified surrogate models or reduced-order approximations to enable real-time inference. Hybrid approaches that integrate both methods, such as using offline-trained physics-informed neural networks (PINNs) for rapid online inference, offer promising solutions. Additionally, ensemble-based approaches, such as particle filters or ensemble Kalman filters, can bridge the gap between the two paradigms by maintaining a probabilistic representation of the parameters while allowing for adaptive updates. A simple comparison between the online and offline approaches are given in Table 6.3.

Ultimately, the choice between offline and online behavior matching depends on the application context: offline approaches excel in high-precision, retrospective analysis, while online methods are indispensable for real-time decision-making and control. Future research directions include improving uncertainty quantification in both paradigms, leveraging SciML architectures like Fourier Neural Operators (FNOs) for efficient surrogate modeling, and exploring multi-fidelity approaches that dynamically switch between offline and online estimation strategies based on computational resources and data availability.

6.4 THE MOABS/DT PLATFORM

To create a Digital Twin for modeling an emission source in environmental sensing applications, such as a point source emission, the model needs to be efficient enough to be computed in real-time (or near real-time), and the model needs to be capable of capturing the key characteristics and relevant dynamics of the physical system. In this book, we aim to capture these goals by utilizing a hybrid DT model that runs in near-real time. We illustrate this using the Methane Odor Abatement Simulator Digital Twin (MOABS/DT).

To understand the MOABS/DT model for a point source, we must start with a proposed model from first principles without a physical system, i.e. simple L1 DT. The governing equations of a trace gas in an advecting field can be represented by a parabolic partial differential equation, namely, the advection diffusion equation (ADE), given by,

$$\partial_t y = \nabla \cdot (D\nabla y) - \mathbf{u}\nabla y + \text{source}, \qquad (6.40)$$

where the y represents the concentration of the gas, D is the diffusion coefficient, and $\mathbf{u} = [u_1, u_2]^T$ is the wind vector field. The source can be represented using Kronecker delta function $\delta_s = \delta(\mathbf{x} - \mathbf{x}_s)$, where $\mathbf{x}_s = [x_{s1}, x_{s2}, x_{s3}]^T$ is the location of the point source. The wind field parameters can be solved generally by using the something like the incompressible Navier-Stokes equation,

$$\partial_t \mathbf{u} + \mathbf{u} \cdot (\nabla \mathbf{u}) - \nu \nabla^2 \mathbf{u} = -\nabla p/\rho + \mathbf{g}. \qquad (6.41)$$

Here $\nu = \mu/\rho$ is the kinematic viscosity, p is the pressure, ρ is the fluid density, and $\mathbf{g}$ is the gravitational acceleration. Solving these systems consecutively can be expensive and thus, some assumptions need to be applied to simplify the models. One such approach to solve this problem was proposed by Farrell et al [5] and is summarized here for completeness. The first assumption is to break the concentration dynamics into discrete filaments (or packets of molecules) based on different length scales (i.e. large scale advection (a), intermediate scale turbulent mixing and stirring (m), and local small scale diffusion (d)). The transport of the i-th filament can then be described by, $\dot{\mathbf{z}}_i$, which is a function of $\mathbf{u}(\mathbf{x}_i)$. One way to write these different length scales is by letting $\dot{\mathbf{z}}_i = \mathbf{u}_a(\mathbf{z}_i) + \mathbf{u}_m(\mathbf{z}_i) + \mathbf{u}_d(\mathbf{z}_i)$. However, the small scale diffusion term can be readily absorbed into the shape of the filament, such that the

TABLE 6.3 Generalized comparison between offline and online behavior matching techniques.

Forward Model Type	Offline Methods	Pros	Limitations	Online Methods	Pros	Limitations
Numerical (PDE-based)	Adjoint optimization, Bayesian inference	High accuracy, full-field parameter recovery	Computationally expensive, requires full dataset	EnKF, 4D-Var, RLS	Real-time updates, can handle time-varying sources	Requires model simplifications, sensitive to noise
Stochastic	EM, MCMC	Probabilistic estimation, captures uncertainty	Requires many samples, slow for high dimensions	Particle filters, Kalman filtering	Adapts to dynamic sources	Computational cost, risk of divergence
Analytic Parameterized	Gradient descent, least squares	Fast convergence, interpretable models	Limited flexibility, requires closed-form solutions	Data assimilation, recursive estimation	Low computational cost	Assumes model is correct
Machine Learning	PINNs, FNOs	Handles high-dimensional systems, learns complex patterns	Requires extensive training data	Online PINNs, RNNs	Adapts to new data, captures temporal changes	Model drift, data dependence

filament's i-th location can be described as,

$$\dot{\mathbf{z}}_i = \mathbf{u}_a(\mathbf{z}_i) + \mathbf{u}_{m_i}(\mathbf{z}_i). \tag{6.42}$$

Now the only thing left to resolve is the wind vector field. One approach is to look at average wind vector such that $\mathbf{u} = \bar{\mathbf{u}} + \mathbf{u}'$, where the overbar $(\bar{\cdot})$ represents the average and the prime $(')$ denotes the deviation from the average. Substituting the decomposition of the wind vector into (6.41) and using the following assumptions, the Navier-Stokes equation can be drastically simplified:

1. Coriolis forces, geostrophic winds, and molecular viscosity can be deemed small and neglected;

2. Measurements are conducted close to the ground where winds are relatively constant w.r.t. the altitude.

The wind field can then be represented by the following equations that resembles the viscous Burgers equation,

$$\frac{\partial \bar{\mathbf{u}}}{\partial t} = -\bar{\mathbf{u}} \cdot \nabla \bar{\mathbf{u}} + \mathbf{K} \cdot \nabla^2 \bar{\mathbf{u}}. \tag{6.43}$$

This is realized by using the simplest K-closure method, which relates the diffusivity $(K_{1,2})$ to, $\overline{u_1' u_1'} = -K_1 \frac{\partial \bar{u}}{\partial x_1}$ and $\overline{u_1' u_2'} = \overline{u_2' u_1'} = -\frac{1}{2}(K_1 \frac{\partial \overline{u_2}}{\partial x_1} + K_2 \frac{\partial \overline{u_1}}{\partial x_2})$ and similarly for the $\overline{u_2' u_2'}$. For the simulation aspect, the wind field meandering can be produced by perturbing the flow with colored noise using a second-order transfer function approach, $H(s) = Ga/(s^2 + bs + a)$ driven by white noise [5] where G is the gain, and a and b are parameters that affect the natural frequency and damping ratio. This colored noise is updated on the corners of the finite grid and adjacent boundary conditions are linearly interpolated (see Fig. 6.3). The meandering behavior (see Fig. 6.2) can be adjusted by scaling a, b, G, and $K_{x,y}$. The interior nodes can be solved implicitly [20] or explicitly by finite differences [23].

The large scale advective term, $\mathbf{u}_a$, can be calculated by using a bilinear interpolation between the nearby wind vectors points or estimated using the nearest neighboring point relative to the filament position, $\mathbf{z}_i$. The grid size used to calculate the the wind field can remain course (large Δx_1 and Δx_2) as it reflects the larger length scales of the model. Additionally, the large spatial separation between points helps to increase simulation runtime. To maintain efficiency we used the nearest neighboring point for this MOABS/DT platform. To calculate the intermediate scale term, $\mathbf{u}_m(\mathbf{z}_i)$, we utilize a random process satisfying, $\dot{\mathbf{u}}_{m_i} = A'\mathbf{u}_{m_i} + B'\eta$ and $\mathbf{v}_m = C'\omega + D'\eta$, where $\omega \in R^n$ and η is a white noise random process with spectral density σ_η^2. The matrices A', B', C', and D' are appropriately sized. The resulting standard deviation grows as $G\sigma_\eta\sqrt{t}$. In the simplest 2D case $D' = \mathbf{0}$, $(B')^T = C' = [1,1]$, and $A' = I$, where I is the identity matrix. To calculate the small scale diffusion, $\mathbf{u}_d$ the use of a growth term, $R(t)$, is needed. This growth term represents the standard deviation of the packet of molecules inside the discrete filament,

$$\frac{dR_i}{dt} = \frac{\gamma}{2R_i} \quad \rightarrow \quad R_i(t) = (R_{i0}^2 + \gamma t)^{1/2}, \tag{6.44}$$

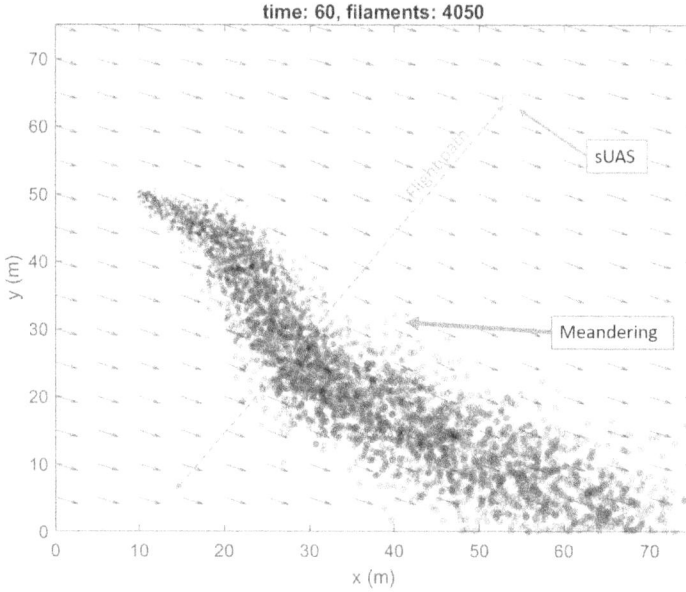

FIGURE 6.2 MATLAB example of the digital twin with meandering. (reused from [9])

where $r_i(\mathbf{x}(t)) = ||\mathbf{x}(t) - \mathbf{z}_i(t)||$ is the Euclidean distance from the sensor at $\mathbf{x}(t)$ and the i-th filament, $R_{i0} = R_i(0)$ and $\gamma > 0$ is a growth rate. This allows for the concentration of individual molecules in a filament to be calculated by,

$$C_i(\mathbf{x}(t), t) = \frac{Q}{\sqrt{8\pi^3 R_i^3(t)}} \exp\left(-\frac{r_i^2(\mathbf{x}(t))}{R_i^2(t)}\right), \qquad (6.45)$$

where Q is the filament source rate in molecules per filament. The total concentration measured, $C(\mathbf{x}, t)$, can be calculated given an effective sensor area, A_e, where N represents the number of filaments inside $A_e = \pi r_e^2$.

$$C(\mathbf{x}, t) = \sum_{i=1}^{N} C_i(\mathbf{x}, t) \qquad (6.46)$$

The effective area can be approximated as a circle with radius r_e. The sensor can be then modeled as a low pass filtered system with thresholding, $\dot{y} = f_{BW}(C(\mathbf{x}, t) - y)$. The sensor bandwidth is f_{BW} and threshold is τ_y. The measured concentration, y_m would then be,

$$y_m(\mathbf{x}, t) = \begin{cases} y(\mathbf{x}, t), & \text{if } y(\mathbf{x}, t) > \tau_y \\ 0, & \text{otherwise} \end{cases}. \qquad (6.47)$$

At this point, the current form of the L1 DT can represent the gas plume in a 2D flow field only. In most scenarios, the gas dispersion needs to be represented in a 3D flow field – usually with building resolving capabilities. However, solving the wind field for 3D case becomes increasingly intractable do to the increase in computation as the number of discrete points grows very quickly (e.g. scaling on the order of $1/\Delta x^3$).

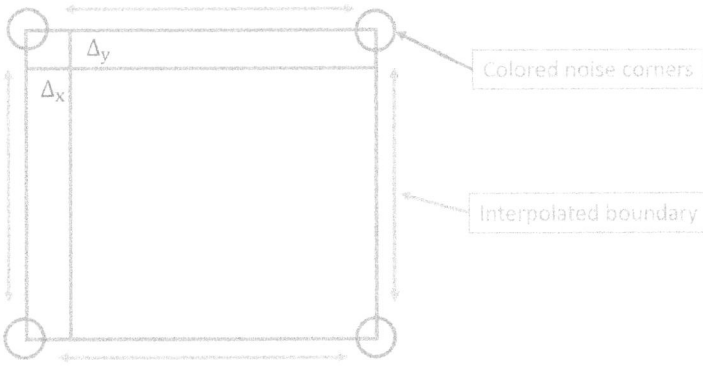

FIGURE 6.3   Scheme for updating the colored noise in the corners interpolating the adjacent boundaries. (reused from [9])

To start, we will examine a rural case with no buildings and a first approximation of a 3D gas dispersion. In [8], the model was extended to 3D by utilizing a detection probability based on experimental observations in [10, 25]. The experiments showed that the detection probability varies as a function of altitude and wind speed. An example of this can be seen in Fig. 5.3, where the sigmoid function can be fit to the probability of detection given different wind conditions or atmospheric stability classes. By using this knowledge, one can essentially extend the 2D model to 3D. Given that sensors do not always detect the target gas, we can introduce a conditional detection probability P_d. Letting the P_d be a function of downwind distance d, altitude x_3, and wind velocity $|\mathbf{u}_a|$. Making the assumption that the lateral or cross wind position does not affect the probability with respect to the simulation, the relation becomes, $P_d(\delta_d|x_3) = f(d, x_3, |\mathbf{u}_a|)$. Given the nature of the mass balance method, the downwind distance d does not readily change, and therefore can be approximated as

$$P_d(\delta_d|x_3) \approx \frac{P_d(\delta)}{1 + e^{M(x_3 - x_{3,bias})}}. \tag{6.48}$$

Here M is the slope at the half probability point, and z_{bias} is the altitude at which half probability occurs. From the sigmoid curves in [10] (see Fig. 5.3), $P_d(\delta) = [0.62, 0.9]$, $M = [0.42, 1.5]$, and $x_{3,bias} = [7.2, 9]$, where the average can be chosen to mirror the conditions of the experiment. However, if the gas is not in contact with the ground, the distribution does not capture the vertical behavior. By using a Gaussian normal distribution,

$$P_d(\delta_d|x_3) \approx P_d(\delta) \exp\left(\frac{-(x_3 - x_{3,bias})^2}{2\sigma_3^2}\right), \tag{6.49}$$

we can alleviate this limitation. Although this only works if the source location and downwind distance are known ahead of time. Additionally, there is a relationship between the value of M, z_{bias}, and the atmospheric stability classes that needs to be further investigated. The atmospheric stability classes (A-F), first introduced by Pasquill in 1961 and later reformulated by Gifford, are typically referred to as the

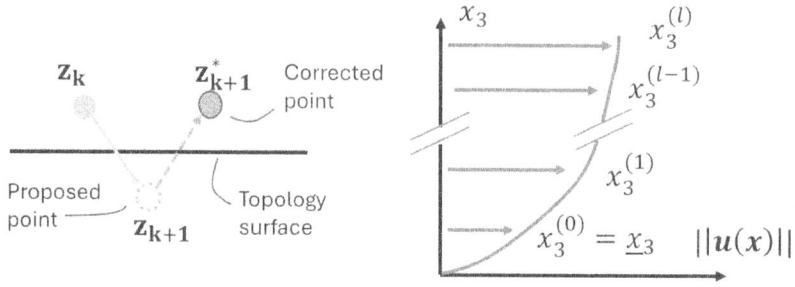

FIGURE 6.4 A diagram depicting (left) the simple collision model, and (right) the power law scaling of a 2D wind field to 3D.

Pasquill-Gifford (PG) curves [12]. The PG curves relate wind speed and solar irradiance to the dispersion coefficients that govern the spread of the plume. This is represented in the Gaussian plume model or sometimes referred to as the infamous 'bell curve' (recall (4.3) from Chapter 4). In [11], the DT model was extended to 3D efficiently by applying a power law scaling (see [22]) of the 2D windfield in the x_3 direction,

$$\mathbf{u}^{(l)}(\mathbf{x}, t) = \underline{\mathbf{u}}(\mathbf{x}, t) \left(\frac{x_3^{(l)}}{\underline{x_3}} \right)^{\alpha} + \epsilon(t), \tag{6.50}$$

where $\underline{\mathbf{u}}(\mathbf{x}, t)$ represents the windfield at ground level (solved implicitly), $\underline{x_3}$ represents the height of the wind sensor (typically can be set to the minimum flight altitude e.g. 2 m), and $\mathbf{u}^{(l)}(\mathbf{x}, t)$ is the l-th 2D windfield at height $x_3^{(l)}$. A normally distributed white Guassian noise, $\epsilon \sim \mathcal{N}(0, \sigma_{u3})$, is introduced to allow for fluctuations in all three principal directions. The choice of α and σ_{u3} can be chosen given the atmospheric stability. The site topology can also be introduced into the DT model by using simple geometric shapes. The interaction between topology and gas filaments can be implemented with a filament collision model. A simple model can be formulated that detects the filament below the topological map, and then reflects the vector based on the normal to the surface. An illustration of the power-law scaling and collision model can be seen in Fig. 6.4.

The simulation source rate, $\hat{Q}_*$, given in molecules per second, and is determined by the number of filaments that are discretely released from the point source location. The particular way the filaments are released depends on two parameters, the number of filaments per puff n_f and the number of puffs per second n_p. The filament source rate can then be defined by $Q = \hat{Q}_*/(n_f n_p)$, which is used in (6.45). The different combinations of these parameters can result in qualitative changes in the measured signal from the sensor. This qualitative behavior can be matched to the measured signal of the physical system and sensor (i.e. behavior matching). The behavior matching can be undertaken given an observable data set that captures these qualitative changes within the model. The key point here is observable data and thus the question becomes, how to get it? The answer depends generally on two things: sensing in the right place and sensing at the right time. Given a set of observations $\mathbf{s}$ we want to infer the states or plume field, $y(\mathbf{x}, t)$. This observation problem can be formulated as a state estimation problem using orthogonal decomposition methods based on singular value decomposition (SVD) [4]. Essentially the problem can be represented as a $\mathcal{L}_2$ minimization,

$$\mathbf{s} = \mathbf{H}\mathbf{y} \approx \mathbf{H}\boldsymbol{\Phi}\nu, \tag{6.51}$$

$$\nu \in \arg\min_{\tilde{\nu}} ||\mathbf{s} - \mathbf{H}\boldsymbol{\Phi}\tilde{\nu}||_2^2, \tag{6.52}$$

where $\boldsymbol{\Phi}$ represents the orthogonal modes and ν the associated coefficients. This approach can yield solutions that are often ill-posed and not necessarily unique. Assuming ν^* can be written as $\nu^* = \boldsymbol{\Phi}^+\mathbf{y}$, it can be shown that the estimation error depends on the approximation basis $\boldsymbol{\Phi}$ (orthogonal modes) as well as the measurement operator, $\mathbf{H}$,

$$||\mathbf{y} - \hat{\mathbf{y}}|| = ||(\mathbf{I} - \boldsymbol{\Phi}(\mathbf{H}\boldsymbol{\Phi})^+\mathbf{H})\mathbf{y}||. \tag{6.53}$$

The symbol ($^+$) represents the Moore-Penrose pseudo-inverse. The measurement operator depends entirely on the sensor location and not the orthogonal modes. Therefore, the choice of sensor location is an important task. For the single sUAS measurement system, this comes in the form of path planning and is subject to disturbances from changes in the stability of the wind. When applied to the mass balance method (recall Section 4.1.3), for instance, the path planning involves scanning back and forth at different altitudes inside a region of interest. The size of this region can incorporate twice the expected plume dispersion (deviation from the plume centerline) σ_2 and σ_3, such as in [10] to ensure a likely encapsulation of the plume (see Fig. 6.5). This approach aims to capture rich signal data for conducting the mass balance method. However, this process takes time to complete with a single sensor system, and thus can present spatial-temporal issues (linked to observability) when reconstructing the mass flux.

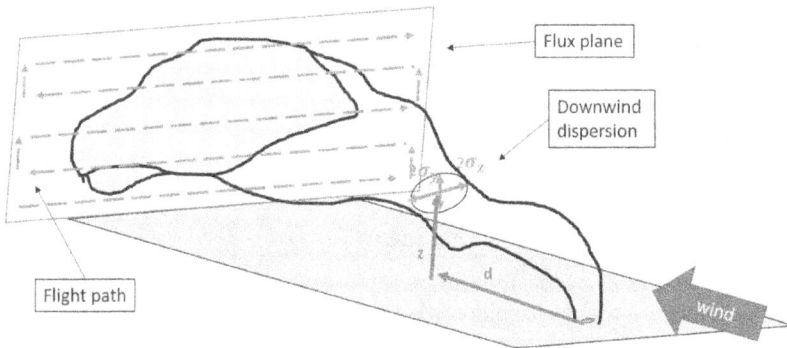

FIGURE 6.5 An example curtain flight path for mass balance calculations. (reused from [9])

To advance the L1 Digital Twin to L2, we need to have the physical twin exist and have the virtual system model of the physical twin (which we just discussed). This step requires a system identification optimization of the L1 DT to estimate the parameters. Given a dataset with observable properties and a L1 DT with tunable parameters,

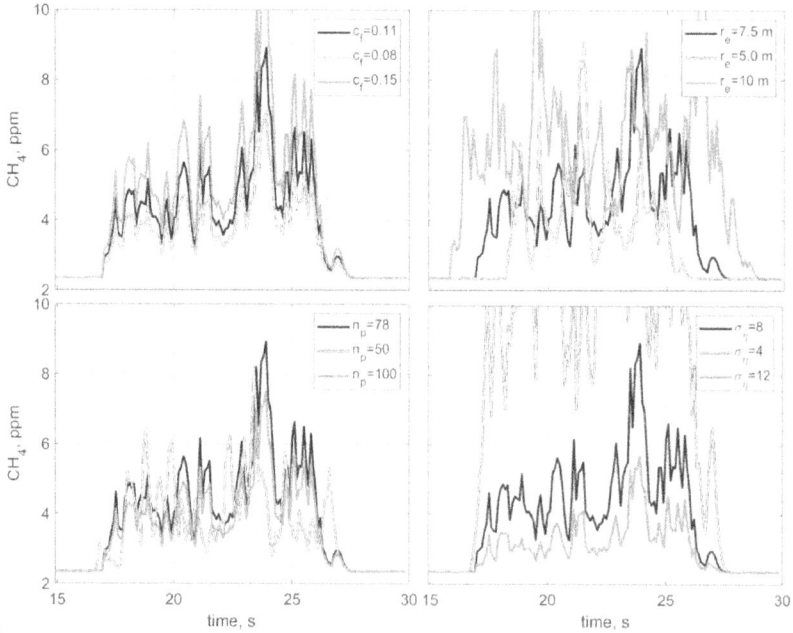

FIGURE 6.6 The timeseries behavior and sensitivity with tuning parameters θ given one pass through the plume. (reused from [9])

minimization on a cost function can be undertaken. The tunable parameters (see Fig. 6.6) in this DT model consist of: the effective radius r_e, the number of filaments per puff n_f, the number of puffs per second n_p, the intermediate scale mixing and stirring dispersion σ_η, and a correction factor c_f where the corrected simulation source rate is $(\hat{Q}_*)_c = c_f \hat{Q}_*$. The value of $\hat{Q}_*$ is chosen to be the same as the physical source rate, such that $Q = (\hat{Q}_*)_c / (n_f n_p)$, and c_f is determined through behavior matching optimization. Once this minimization is complete, the L2 Digital Twin can be used for performance evaluation, method development, etc. For example, it will allow for the testing different path planning strategies to improve mass balance method results.

Pause and Reflect

Given advancements in computation and machine learning, including foundation models, how will the digital twin strategies propel the way modeling and sensing are performed?

6.5 CHAPTER SUMMARY

In this chapter we focused on understanding what a Digital Twin is, how to represent it within several modeling frameworks (i.e. through first principles, stochastic differential equations, parameterized models, machine learning, and hybrid), and how to behavior match the DT with the physical system (i.e. estimate the model parameters) in online and offline settings. We learned about the challenges associated with high fidelity modeling and how that may impact online behavior matching. Lastly, we introduced the L1 MOABS/DT platform for representing an emission source. The model is based on a hybrid formulation, where the wind field is solved using a deterministic framework and the concentration field using a Lagrangian-based framework.

Bibliography

[1] Kaushik Bhattacharya, Bamdad Hosseini, Nikola B Kovachki, and Andrew M Stuart. Model reduction and neural networks for parametric PDEs. *The SMAI Journal of Computational Mathematics*, 7:121–157, 2021.

[2] Stefan Boschert and Roland Rosen. Digital twin—the simulation aspect. In *Mechatronic Futures*, pages 59–74. Springer, 2016.

[3] Yuyao Chen, Lu Lu, George Em Karniadakis, and Luca Dal Negro. Physics-informed neural networks for inverse problems in nano-optics and metamaterials. *Optics Express*, 28(8):11618–11633, 2020.

[4] N Benjamin Erichson, Lionel Mathelin, Zhewei Yao, Steven L Brunton, Michael W Mahoney, and J Nathan Kutz. Shallow neural networks for fluid flow reconstruction with limited sensors. *Proc. of the Royal Society A*, 476(2238):20200097, 2020.

[5] Jay A Farrell, John Murlis, Xuezhu Long, Wei Li, and Ring T Cardé. Filament-based atmospheric dispersion model to achieve short time-scale structure of odor plumes. *Environmental Fluid Mechanics*, 2(1-2):143–169, 2002.

[6] Maryam Farsi, Alireza Daneshkhah, Amin Hosseinian-Far, and Hamid Jahankhani. *Digital Twin Technologies and Smart Cities*. Springer, 2020.

[7] Michael Grieves. Digital twin: manufacturing excellence through virtual factory replication. *White paper*, 1:1–7, 2014.

[8] Derek Hollenbeck and YangQuan Chen. Characterization of ground-to-air emissions with sUAS using a digital twin framework. In *Proc. of the 2020 International Conference on Unmanned Aircraft Systems (ICUAS)*, pages 1162–1166. IEEE, 2020.

[9] Derek Hollenbeck and YangQuan Chen. A digital twin framework for environmental sensing with sUAS. *Journal of Intelligent & Robotic Systems*, 105(1):1, 2022.

[10] Derek Hollenbeck, Moataz Dahabra, Lance E Christensen, and YangQuan Chen. Data quality aware flight mission design for fugitive methane sniffing using fixed wing sUAS. In *Proc. of the 2019 International Conference on Unmanned Aircraft Systems (ICUAS)*, pages 813–818. IEEE, 2019.

[11] Derek Hollenbeck, Demitrius Zulevic, and YangQuan Chen. MOABS/DT: Methane Odor Abatement Simulator with Digital Twins. In *Proceedings of the 2021 IEEE 1st International Conference on Digital Twins and Parallel Intelligence (DTPI)*, pages 378–381. IEEE, 2021.

[12] C Hunter. A recommended Pasquill-Gifford stability classification method for safety basis atmospheric dispersion modeling at SRS. Technical report, Savannah River Site (SRS), 2012.

[13] George Em Karniadakis, Ioannis G Kevrekidis, Lu Lu, Paris Perdikaris, Sifan Wang, and Liu Yang. Physics-informed machine learning. *Nature Reviews Physics*, 3(6):422–440, 2021.

[14] Aditi Krishnapriyan, Amir Gholami, Shandian Zhe, Robert Kirby, and Michael W Mahoney. Characterizing possible failure modes in physics-informed neural networks. *Advances in Neural Information Processing Systems*, 34:26548–26560, 2021.

[15] Zongyi Li, Nikola Kovachki, Kamyar Azizzadenesheli, Burigede Liu, Kaushik Bhattacharya, Andrew Stuart, and Anima Anandkumar. Fourier neural operator for parametric partial differential equations. *arXiv preprint arXiv:2010.08895*, 2020.

[16] Ziming Liu, Varun Madhavan, and Max Tegmark. Machine learning conservation laws from differential equations. *Physical Review E*, 106(4):045307, 2022.

[17] Lu Lu, Pengzhan Jin, Guofei Pang, Zhongqiang Zhang, and George Em Karniadakis. Learning nonlinear operators via DeepONet based on the universal approximation theorem of operators. *Nature Machine Intelligence*, 3(3):218–229, 2021.

[18] Azad M Madni, Carla C Madni, and Scott D Lucero. Leveraging digital twin technology in model-based systems engineering. *Systems*, 7(1):7, 2019.

[19] Guri I Marchuk. *Adjoint Equations and Analysis of Complex Systems*, volume 295. Springer Science & Business Media, 2013.

[20] N Mohamed. Fully implicit scheme for solving Burgers' equation based on finite difference method. *Egyptian Journal for Engineering Sciences and Technology*, 26:38–44, 2018.

[21] Yu Nesterov. Gradient methods for minimizing composite functions. *Mathematical Programming*, 140(1):125–161, 2013.

[22] Jennifer F Newman and Petra M Klein. The impacts of atmospheric stability on the accuracy of wind speed extrapolation methods. *Resources*, 3(1):81–105, 2014.

[23] Richard H Pletcher, John C Tannehill, and Dale Anderson. *Computational Fluid Mechanics and Heat Transfer*. CRC Press, 2012.

[24] Maziar Raissi, Paris Perdikaris, and George E Karniadakis. Physics-informed neural networks: A deep learning framework for solving forward and inverse problems involving nonlinear partial differential equations. *Journal of Computational Physics*, 378:686–707, 2019.

[25] Brendan J Smith, Garrett John, Lance E Christensen, and YangQuan Chen. Fugitive methane leak detection using sUAS and miniature laser spectrometer payload: system, application and groundtruthing tests. In *Proc. of the 2017 International Conference on Unmanned Aircraft Systems (ICUAS)*, pages 369–374. IEEE, 2017.

[26] John M Stockie. The mathematics of atmospheric dispersion modeling. *SIAM Review*, 53(2):349–372, 2011.

[27] Shashank Subramanian, Peter Harrington, Kurt Keutzer, Wahid Bhimji, Dmitriy Morozov, Michael W Mahoney, and Amir Gholami. Towards foundation models for scientific machine learning: Characterizing scaling and transfer behavior. *Advances in Neural Information Processing Systems*, 36:71242–71262, 2023.

[28] Jairo Viola and YangQuan Chen. Digital twin enabled smart control engineering as an industrial AI: A new framework and a case study. *arXiv preprint arXiv:2007.03677*, 2020.

[29] Sifan Wang, Yujun Teng, and Paris Perdikaris. Understanding and mitigating gradient flow pathologies in physics-informed neural networks. *SIAM Journal on Scientific Computing*, 43(5):A3055–A3081, 2021.

[30] Yinhao Zhu and Nicholas Zabaras. Bayesian deep convolutional encoder–decoder networks for surrogate modeling and uncertainty quantification. *Journal of Computational Physics*, 366:415–447, 2018.

CHAPTER 7

Case Studies: Digital Twins

Digital twins are useful tools in modeling complex systems, but more importantly, they can be used to conduct preliminary designs, performance assessments, and more (see Table 6.1). In this chapter, we will take a look at some DT case studies focused on behavior matching, emission quantification method development, and performance assessment.

7.1 BEHAVIOR MATCHING GROUND-TO-AIR EMISSIONS

In the previous chapter, we introduced the MOABS/DT platform and the parameters that need to be tuned in order to behavior match the model with the physical system. This case study looks at one approach to behavior matching, in the offline sense, by looking at the behavior of the measured time series signals compared to the model's estimated time series signals. The challenge is in comparing the two signals which may not 'line up' in the traditional sense of least squares.

7.1.1 Method and Materials

The experimental data used to convert the L1 Digital Twin to L2 was based on the 2018 August controlled release single source field experiment in the Merced Vernal Pools and Grassland Reserve (MVPGR) just north of the University of California, Merced main campus (See experimental setup in Section 5.2.3). The sUAS used was the DJI M210, equipped with the OPLS [6].

7.1.1.1 Behavior Matching Offline Optimization

Once the experimental data is collected, the most data information rich flights are to be selected to perform behavior matching with the L1 digital twin. In this experiment, some of the curtain flights (such as flights 13 and 14) resulted in no detected emissions and cannot be used to behavior match the model (see Table 7.1). After analyzing the 16 flights using the mass balance method with kriging (see Fig. 7.1), flight 1 showed stable wind conditions, consistent methane detection throughout the ascending transects, and good estimation of the true source rate. Thus, flight 1 was chosen for behavior matching the sensor signal of the L1 digital twin to the experimental

DOI: 10.1201/9781003669470-7

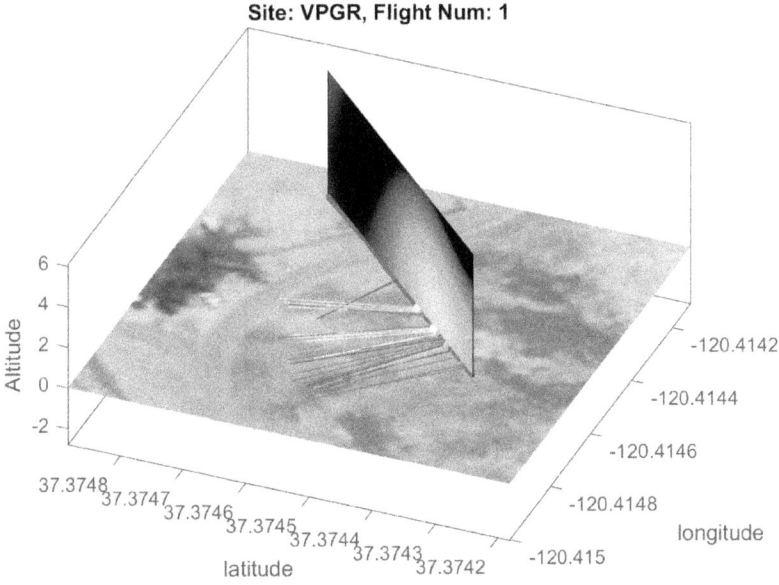

FIGURE 7.1 Example mass balance measurement for one VFP curtain flight. (reused from [14])

sensor signal (see Fig. 7.2). The simulation source rate $\hat{Q}_*$ was set to ($\hat{Q}_* = 2.4 \times 10^{24}$ molecules/s). Using the PG curves given the current meteorological conditions (i.e. atmospheric stability) the initial condition of σ_η was chosen to be 8 m. The wind speed and direction measured from the experiment was used as a direct input into the L1 digital twin, using the mean wind field approximation. Using the behavioral characteristics of the hyperparameters shown in Section 6.4 and in an attempt to reduce complexity, let $n_f = 1$, such that $n_f n_p$ has units as filaments per second. Heuristically tuning the hyperparameters, to give a best guess of $\theta = [c_f, r_e, n_p, \sigma_\eta]$, the following initial conditions were chosen: $c_f = 0.2$, $r_e = 5$ m, $n_p = 40$ puffs/sec., and $\sigma_\eta = 8$ m. It can be also seen that there is a lag present in the system (see Fig. 7.2) which can be attributed to model inconsistencies as well as initial conditions of the plume. Since in this case study there is currently no implementation of plume rise (see MOABS/DT in Section 6.4), the average altitude where concentration detection occurred was used to determine the $x_{3,bias}$ term and the vertical dispersion was set to $\sigma_3 = 5$ m.

If we represent the L1 digital twin model as, $\mathcal{M}_{DT}(\cdot)$, with hyperparameters θ as a function at time t, from a starting time t_0 to the ending time t_f, the resulting data set can be described as

$$\{\hat{y}_t, \mathcal{Z}_t\}_{t=t_0}^{t=t_f} = \{\mathcal{M}_{DT}(\mathbf{x}(t), \mathbf{u}(t)|\theta)\}_{t=t_0}^{t=t_f}, \tag{7.1}$$

with the sUAS position (i.e. sensor position) $\mathbf{x}$ and $\mathbf{u}$ describes the wind speed and direction measurement using a model 81000 3D ultrasonic anemometer from the R. M. Young company, USA. The output at the k-th time step consists of a

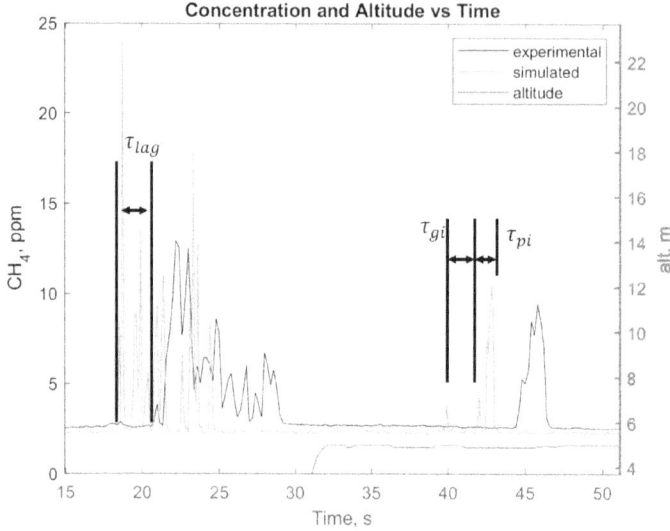

FIGURE 7.2 The concentration time series in ppm during flight curtain 1, where $n_p = 40$ puffs/sec and the symbols τ_{gi}, τ_{pi}, and τ_{lag} represent the i-th gap length, i-th pulse length, and the time lag, respectively, between the simulation and experimental data [12]. (reused from [14])

concentration measurement $\hat{y}_k = \hat{y}(\mathbf{x}(t_k), t_k)$ and filament positions $\mathcal{Z}_k = \{\mathbf{z}_i^{(k)}\}_{i=1}^{N_f}$. In the physical experiment the locations of the plume filaments are not observable given that methane gas is not visible to the human eye. The plume can potentially be observed using a large number of high quality sensors, of which, is expensive, difficult to set up, and only be inferred using spatial interpolation techniques. Thus, the only output available to optimize over is the n_t concentration measurements, $\hat{y}(\mathbf{x}, t)$, in the least squares sense. In order to compensate for the stochastic behavior of the plume dynamics, the cost function was averaged over $n_T = 20$ trials. Unfortunately, this leads to multiple possible solutions to the minimization. Therefore, additional constraint was placed on the hyperparameter space, Θ, and a regularization term can be added to constrain the magnitude of the concentration measurement such that

$$\hat{\theta} = \arg\min_{\theta \in \Theta} J(\theta), \quad \Theta = \{\theta | \theta_{min} \leq \theta \leq \theta_{max}\}, \tag{7.2}$$

$$J(\theta) = \frac{1}{n_T} \sum_{k=1}^{n_T} \left[0.2(\sum_{i=1}^{n_t} y_i - \sum_{j=1}^{n_t} \hat{y}_j^{(k)})^2 + 0.8(y_{max} - \hat{y}_{max}^{(k)})^2 \right], \tag{7.3}$$

where $y_{max} = \max(\mathbf{y})$ is the maximum observed measurement value from all the observations in the time series, and conversely $\hat{y}_{max}^{(k)}$ in the k-th trial. The cost function was designed in such a way to limit the amount of signal energy in the time series result but also constrain the shape to more closely match the peak of the observed signal. The weights were chosen to balance the effects of each term and were determined by trial and error.

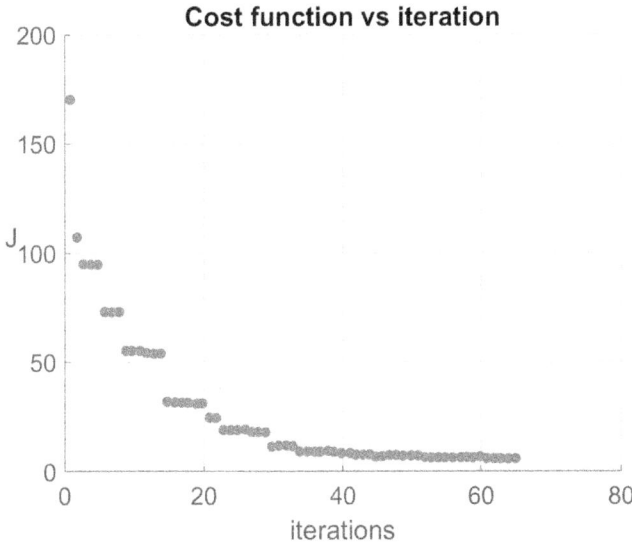

FIGURE 7.3 Pattern search method convergence plot for behavior matching of flight 1 given the cost function, $J(\theta)$. The algorithm converged as mesh size tolerance was reached to ≈ 5.813. (reused from [14])

The solution to (7.2) can be found using a variety of optimization methods and techniques, such as Gradient Descent or Stochastic Gradient Descent. Faster convergence can be achieved using Hessian-based approaches but due to the cost of the forward model calculation and considering complexity of implementation these methods were not considered. Other gradient-free approaches are more desirable for this problem such as Genetic Algorithm, Extremum Seeking Control, or Pattern (Direct) Search can be applied. With simplicity in mind, the Pattern Search Optimization was used here to behavior match the digital twin to the experimental data. After 65 iterations and over 390 cost function evaluations (7880 forward model runs), the gradient of the cost function $dJ/dt \approx 0$ (see Fig. 7.3). The model parameters were found to be: $c_f = 0.11$, $r_e = 7.5$ m, $n_p = 78$ puffs/s, and $\sigma_\eta = 8$ m.

7.1.1.2 Quantification Methods

To calculate and perform the mass balance method, sometimes called the box method, requires a way to measure the flux entering or leaving a control volume (see Section 4.1.3). Typically, upwind and downwind curtain flights are used to measure the flux. However, in the case of a controlled release the downwind curtain flight is sufficient to calculate the flux, granted the plume is encapsulated in the flight path (see Fig. 6.5). Using the time series vector $\hat{\mathbf{y}} = [\hat{y}(\mathbf{x}(t_0), t_0), \ldots, \hat{y}(\mathbf{x}(t_{n_t}), t_{n_t})]^T$ with the corresponding measurement locations, $\{\mathbf{x}(t_k)\}_{k=0}^{k=n_t}$, a concentration matrix, $\mathbf{Y}$ can be estimated on an n by m spatial grid through spatial interpolation methods. The flux calculation can then be given as the integration over curtain area with the average wind vector multiplied by the cosine angle of the normal vector to the curtain with the enhanced concentration,

$$\hat{Q} = \int_A \overline{\mathbf{u}} \cdot \hat{\mathbf{n}}[(\mathbf{Y} - y_b)]dA. \tag{7.4}$$

Here y_b represents the background concentration measurement. An example of a flux plane calculation can be shown in Fig. 7.1. Several kinds of techniques can be applied to the time series data to generate $\mathbf{Y}$, such as Ordinary Kriging, Inverse Distance Weighting (IDW), or statistical gas distribution modeling (e.g. Kernel DM+V/ W). For the case of Kriging, a linear unbiased estimator can be solved using a variogram (or semi-variogram – see Section 4.1.3).

Generally, the redundancy and closeness covariance matrix are calculated from fitting the experimental variogram to a function with respect to the spatial distance or lag, such as the exponential or Gaussian, and knowing the covariance of the measured points (referred to as the sill). This method requires second-order stationarity in the spatial observations [29]. Unfortunately this is only approximately the case during stable atmospheric conditions. Methods such as IDW look at the inverse distance when determining weights and does not consider other spatial statistics as in Kriging. There has been some efforts to improve this by looking at minimum error variance within IDW [3]. Wind information has been incorporated into Kernel DM+V/W approach in 3D [23]. Another adaptation to statistical gas distribution mapping approach was implemented in [11] by using the Gaussian plume kernel, which incorporates the wind information into the covariance function in determining the weights. The downside to this approach is that it assumes the source location. This is not ideal for determining the flux in this case. The spatial-temporal effects on modeling have been looked at as a separation between the spatial covariance and the exponential decay between temporal measurements [2, 31],

$$w_i^{(k)} = \mathcal{N}(|\{\mathbf{x}\}_i - \mathbf{x}^{(k)}|, \sigma)\varphi(t_*, t_i), \tag{7.5}$$

where $\mathcal{N}$ is a Gaussian weight function, $(\cdot)^{(k)}$ represents the k-th grid point, $\{\mathbf{x}\}_i$ are the sensor measurements at time t_i, and φ is given as,

$$\varphi(t_*, t_i) = e^{-a_\varphi(t_* - t_i)}. \tag{7.6}$$

However, when dealing with complex systems such as in turbulent environments there may be several factors that determine what the value of a_φ is. If we look at the weight summation of exponential terms, with elements increasing toward infinity, it can be described by fractional-order dynamic behavior [22]. The Mittag-Leffler function [26] can be used to describe this behavior

$$\mathcal{E}_{\alpha,\beta}(t) = \sum_{k=1}^{\infty} \frac{t^k}{\Gamma(\alpha k + \beta)}, \tag{7.7}$$

such that when $\alpha = \beta = 1$, $\mathcal{E}_{1,1}(t) = e^t$. Given that $\beta = 1$ and letting α decrease we can see in Fig. 7.4 that the tail behavior decays much more slowly than the

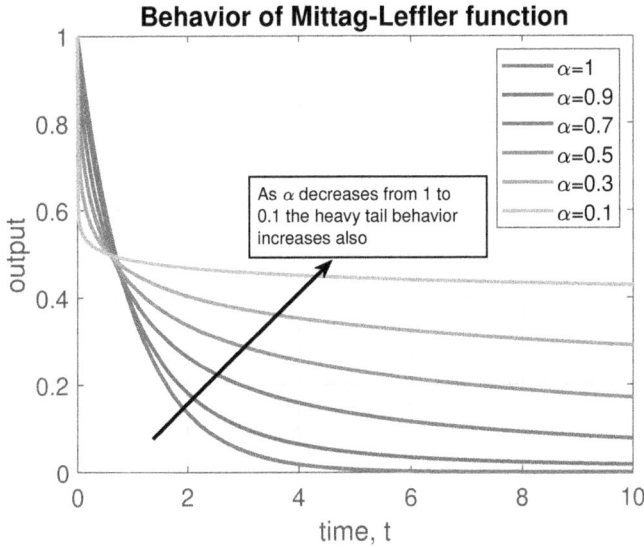

FIGURE 7.4 Response of the single parameter Mittag-Leffler function for $\mathcal{E}_\alpha(-t^\alpha)$ as $\alpha \in [0, 1]$. (reused from [14])

exponential function. By combining the approach in (7.5) with (7.7) we can apply a new weighting scheme with tuning parameter, α,

$$w_i^{(k)} = \mathcal{N}(|x_i - x^{(k)}|, \sigma)\mathcal{E}_{\alpha,1}(-a_\varphi|t_* - t_i|^\alpha). \tag{7.8}$$

The value of t_* can be chosen for each grid point based on the time t_i of the closest measured point that minimizes $|\{\mathbf{x}\}_i - \mathbf{x}^{(k)}|$. For the single sensor mapping problem we test the change in sensitivity for $\alpha = 1$, 0.5, and 0.1.

7.1.2 Results and Discussion

After collecting 10 digital twin simulation outputs using the experimental position, concentration observations and wind measurements for each flight, spatial interpolation was carried out using Kriging, IDW, and Kernel DM/ V to see a comparison in quantitative results across the different approaches. It can be observed in Fig. 7.5 that all the spatial interpolation methods tend to perform with respect to the quantification performance. The proposed spatial-temporal method using (7.8), also does not show noticeable performance improvement. Therefore, there is still future work to be done to address the optimal configurations of these methods and their hyperparameters, as well as, how the spatial-temporal measurement trajectory plays a role with estimation (e.g. affecting observability). It can further be observed that the source rate estimations fluctuate in value. These fluctuations seem to depend on weather conditions and turbulence (see Table 7.1). Given small deviations in the wind, the sUAS curtain flight path can appropriately capture the spatial data through the mass balance plane. When the wind deviates slightly, the data we capture can sometimes be spread out over the flux plane (i.e. like a blurred panoramic image). As a result, this can lead to higher source rate estimations. Conversely, if the meandering happens in the right way, poor detection and characterization of the plume will occur,

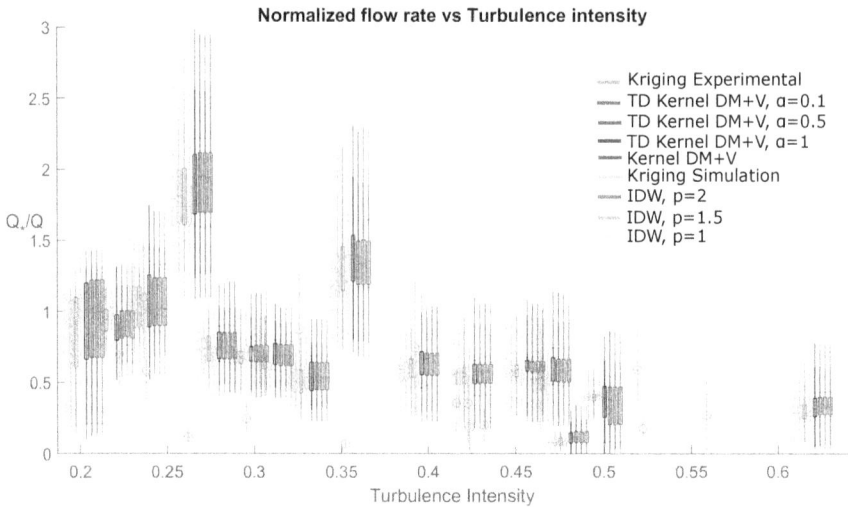

FIGURE 7.5   Here the turbulence intensity, $T_i = \sigma_u/\bar{u}$, of the wind is plotted against the normalized flow rate, which trends toward $\hat{Q}_*/Q = 1$ as $T_i \to 0$. A boxplot of the source estimation for each flight (set of 10 simulations), given a spatial interpolation method, is shown for each T_i condition observed from the experimental data. A smaller boxplot (shown within each boxplot) contains tails that represent the uncertainty. We can observe 1) several flights agree with experimental data 2) majority of flights underestimate the flow rate in short due to increased T_i and 3) the remaining flights that disagree are likely not capturing vertical plume behavior.(reused from [14])

resulting in under-estimation of the source rate. Since the wind direction fluctuations correlate to different stability classes, it makes sense that there should be some correlation with the turbulence intensity as well (see Fig. 7.5). Thus, mindfulness of the current atmospheric conditions is important when planning and conducting mass balance flights. Furthermore, the wind behavior, to some extent, is always changing in magnitude and shifting in direction. It seems this behavior is very rarely calm and stable, which may put physical limitations on the sensitivity of accurate spatial sampling with single mobile sensor systems. Comparing the experimental data to that of the digital twin one can observe qualitative similarities in the distribution of sensor detections, as seen in Fig. 7.6a. Additionally, larger less frequent spikes at lower altitudes were observed that are consistent with what was experimentally observed (see Fig. 7.6b).

7.2   EMISSION QUANTIFICATION WITH MULTI-sUAS DE-CVT

Up until now, we have only considered one sUAS while performing detection, localization, or quantification tasks. Typically, in order to quantify the emission source, a vertical flux plan flight path needs to be flown in order to estimate the source rate once. Then successive flights need to be conducted to get an average source estimation

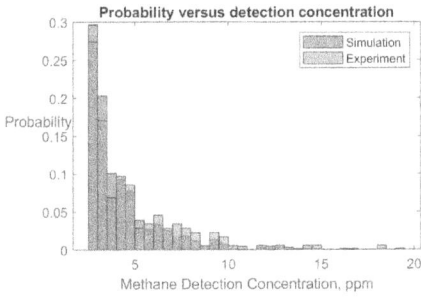

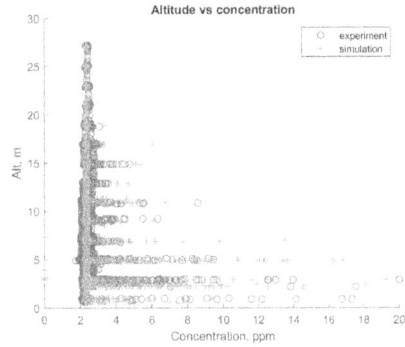

(a)

(b)

FIGURE 7.6 (a) The detection concentrations from the simulation and experimental sensor are compared in a probabilistic histogram. While the distributions show that the majority of indications are near background levels, they also agree with the less likely larger concentrations detected in the experiment. (b) The detection altitude versus measured concentrations from the simulation and experimental sensor are compared. The distributions match well with experiment and the probability of detection declines rapidly as the altitude increases from 10 to 15 m. (reused from [14])

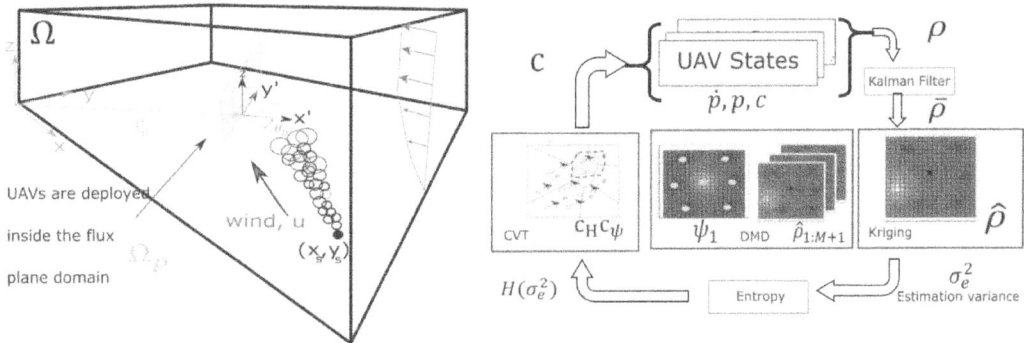

FIGURE 7.7 This diagram depicts the simulation environment setup and framework implementation flowchart. (reused from [13] with permission from ©IEEE)

along with the variance. Often, these flights are subject to variations in the weather, leading to spatial temporal estimation issues. *What if multiple sUAS can be used to improve the spatial temporal issues seen with one sUAS?* In this section we discuss the multi-sUAS continuous source rate estimation method based on CVT coverage control, DMD, and the mass balance flux plane with spatial interpolation (outlined in Sections 4.1.3 and 4.2).

7.2.1 Method and Materials

In this framework, we seek to make continuous source rate estimation by controlling the sUAS positions inside a flux plane Ω_p using CVT coverage control, utilize spatial

TABLE 7.1 Results from mass balance method for 16 curtain flights are shown below along with the average experimental wind speed, $\bar{u}_e$ and the $\cos\beta_e$ between the normal vector of the flux plane. The flow rates (experimental Q_* and simulation $\hat{Q}_*$) are in SCFH, the subscripts denote the spatial interpolation method, and the superscripts are the associated parameters, where K is Kriging, IDW is inverse distance weighting, and KDM is Kernel DM+V.

Flt	Q_*	$(\hat{Q}_*)_K$	$(\hat{Q}_*)^{p=2}_{IDW}$	$(\hat{Q}_*)^{p=1}_{IDW}$	$(\hat{Q}_*)_{KDM}$	$(\hat{Q}_*)^{\alpha=1}_{KDM}$	$(\hat{Q}_*)^{\alpha=0.1}_{KDM}$	$\bar{u}_e$	$\cos\beta_e$
1	14.2±3.9	13.0±2.8	12.1±2.6	12.4±2.6	12.0±3.2	12.2±3.2	12.2±3.2	1.82±0.43	0.83
2	11.3±3.7	7.0±2.3	6.5±2.2	6.9±2.3	7.0±2.7	7.0±2.6	7.0±2.6	1.75±0.57	0.96
3	2.4±1.5	1.0±0.7	1.1±0.8	1.1±0.8	1.5±1.3	1.5±1.4	1.5±1.4	1.18±0.50	0.67
4	9.6±4.3	6.6±3.0	7.8±3.5	7.3±3.3	8.2±4.2	8.1±4.1	8.1±4.1	1.30±0.51	0.87
5	6.8±3.4	6.2±3.1	6.9±3.4	6.7±3.3	7.6±4.2	7.6±4.1	7.6±4.1	1.25±0.58	0.91
6	7.5±1.8	12.8±3.1	13.3±3.2	13.2±3.2	13.8±4.5	13.8±4.2	13.8±4.3	2.27±0.54	0.97
7	4.6±2.1	6.7±3.3	7.5±3.7	7.5±3.6	8.0±4.4	7.9±4.3	7.9±4.3	1.37±0.57	0.67
8	3.2±1.0	9.2±3.0	8.7±2.8	9.1±2.9	9.0±3.2	9.1±3.2	9.0±3.2	1.93±0.57	0.93
9	4.6±2.5	6.3±3.5	7.1±3.9	6.9±3.8	7.2±4.3	7.3±4.3	7.3±4.3	1.49±0.63	0.83
10	3.5±2.2	4.1±2.7	3.9±2.5	3.9±2.6	4.2±3.3	4.4±3.3	4.3±3.3	1.45±0.81	0.86
11	2.6±1.0	8.4±2.2	9.5±2.5	9.3±2.4	9.8±3.3	9.8±3.2	9.8±3.2	1.25±0.54	0.52
12	2.3±1.0	7.5±2.0	8.5±2.3	8.4±2.3	9.0±2.8	8.9±2.7	8.9±2.7	0.90±0.47	0.43
13*	0.9±0.3	15.6±5.4	16.8±5.8	15.7±5.4	17.7±6.9	17.3±6.8	17.5±6.8	2.47±0.87	0.98
14*	1.5±0.4	22.5±6.0	23.3±6.2	22.9±6.1	24.4±7.8	24.6±7.6	24.6±7.6	2.52±0.66	0.94
15	9.0±2.0	13.2±2.9	12.1±2.7	13.2±2.9	11.4±3.4	11.7±3.2	11.8±3.3	2.07±0.46	0.96
16	7.6±2.6	5.4±1.9	5.1±1.7	5.3±1.8	4.7±3.1	4.4±3.2	4.5±3.2	0.52±0.27	0.27

interpolation with N sparse measurements to build a concentration map estimation, and then apply the mass balance method [7] to quantify the source rate (see Fig 7.7). However, due to the turbulent behavior the measurements made, they tend to be noisy and intermittent. The severity of the disturbances can be attributed to the current atmospheric stability, which is a function of wind speed and solar irradiance [18]. During more stable conditions (e.g. turbulence intensity is near 0) the data can be considered quasi-stationary (in the flux plane frame of reference). To help smooth this incoming data, a Kalman filter [27, 30] can be applied to the i-th sUAS measurement at time k to obtain $\overline{p}_{k,i}$.

Once the N sUAS flux plane measurements $\overline{p}_k$ are made, an estimation of the $n \times m$ point spatial concentration map $\hat{\rho}_k \in \Omega_p \subseteq \Omega \in \mathbb{R}^3$ can be computed using ordinary kriging, $\overline{p}_k \to \hat{\rho}_k$ in (4.66) from Section 4.2. The exponential semivariogram [29] was used to find $\hat{\rho}_k$. Since the concentration map is assumed to be quasi-stationary, the results from ordinary kriging can be noisy. Therefore we chose to explore the slow moving spatial temporal dynamics of the estimated concentration map inside the domain Ω_p using DMD. Given $M + 1$ snapshots of $\hat{\rho}_k$, the DMD modes, ψ_1, can be calculated from (4.77) to (4.81) and used to represent the density in the CVT cost, $J(P)$,

$$J(P) = \sum_{i=1}^{n} \int_{V_i} [\psi_1(q) + H(\sigma_e^2(q))]|q - p_i|^2 dq,$$

$$\text{for } q \in \Omega_p.$$

(7.9)

In addition to computing the DMD modes, a by-product of the ordinary kriging is the estimation error variance, σ_e^2. This can in turn be used to calculate the entropy of the error variance as a weight for the CVT algorithm. In adaptive sampling, the entropy of information has shown to be better performing in reconstruction efforts [1]. An advantage of using CVT in this case is the automatic distribution of measurement points (e.g. sUASs). This can prevent uncontrolled clustering, which was an issue encountered in [1]. The entropy is defined in the information theory sense as

$$H(\sigma_e^2) = -\sigma_e^2 \ln \sigma_e^2,$$

(7.10)

where σ_e^2 is a normalized quantity. The introduction of the entropy produces a trade-off between exploration (entropy) and exploitation (density). The density-based CVT on one hand, seems to exploit more often the high concentration regions depending on how the sUASs are deployed. For instance, if deployed in a region with multiple constant hotspots, the sUASs take time to expand and cover the total area (see Fig. 7.8). When entropy is included, the sUASs recognize there are areas of uncertainty not explored and therefore add weight to the unexplored area. This entropy weight helps the CVT algorithm explore in a much quicker way, which is ideal for plumes that may meander back and forth. The estimation error as a result of including entropy dropped significantly given this scenario (see Fig. 7.9). The sUAS position update law was a modification to Lloyds by adding entropy. The update is defined as

$$\dot{p}_i = -k_p(p_i - c_{\psi,i}) - k_e(p_i - c_{H,i}),$$

(7.11)

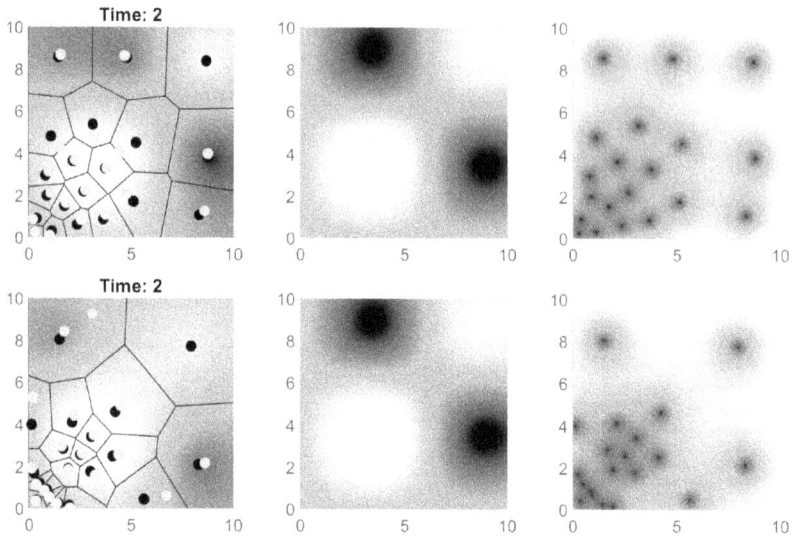

FIGURE 7.8 These plots show the current position/density reconstruction (left), the ground truth (middle), and the entropy (right) of the DE-CVT (top row) and D-CVT (bottom row) strategies for flux reconstruction. (reused from [13] with permission from ©IEEE)

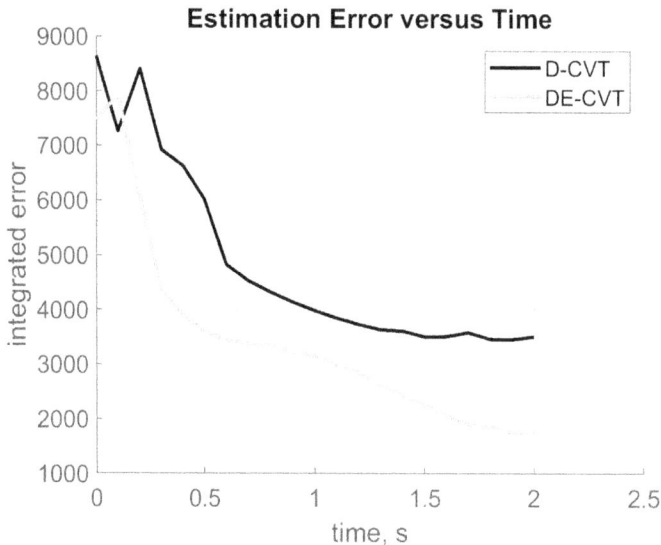

FIGURE 7.9 Given the 2D static case, the estimation error is reduced more with the DE-CVT over the same period of time. This is desirable behavior since the field may change with time in many practical applications. (reused from [13] with permission from ©IEEE)

where k_p and k_e are positive constants used to adjust the weights of density and entropy. The centroids $c_{\psi,i}$ and $c_{\psi,i}$ for the i-th sUAS are solved using,

$$m_{\psi,i} = \int_{v_i} \psi_1(q,t)dq, \quad c_{\psi,i} = \frac{1}{m_{\psi,i}} \int_{v_i} q\psi_1(q,t)dq, \tag{7.12}$$

$$m_{H,i} = \int_{v_i} H(\sigma_e)(q,t)dq, \quad c_{H,i} = \frac{1}{m_{H,i}} \int_{v_i} qH(\sigma_e)(q,t)dq. \tag{7.13}$$

To get the continuous mass flux measurement, a n_a point moving window average is used on the calculated source rate with the first DMD mode as an input

$$\dot{\mathbb{M}}_k = \frac{k_{cf}}{n_a} \sum_{j=k-n_a}^{k} \int_\Omega \bar{u} \cdot \hat{n} \psi_{1,j} \, dA. \tag{7.14}$$

The variable k_{cf} is the conversion factor that converts to desirable units (e.g. SCFH or kg/s), $\bar{u}$ is the average wind vector, and $\hat{n}$ is the normal vector of the flux plane. The moving average was implemented to help smooth out the spiky instantaneous flux measurements (see Fig. 7.11).

In some cases the meandering can shift the plume outside of Ω_p resulting in loss of plume data. To combat this, a simple adaptive algorithm is constructed that relates to the size and position of the domain Ω_p. The domain is defined by the plane position vector $\mathbf{r}_p$, the plane angle θ_p, the maximum height z'_{max}, and maximum width y'_{max}. Given the center of the domain c_Ω and the center of mass of density c_ψ, the flux plane location and domain size can be adjusted adaptively,

$$\begin{bmatrix} \dot{y}'_{max} \\ \dot{z}'_{max} \end{bmatrix} = \begin{bmatrix} -k_y(c_{\Omega,y} - c_{\psi,y}) \\ -k_z(c_{\Omega,z} - c_{\psi,z}) \end{bmatrix}, \tag{7.15}$$

$$\begin{bmatrix} \dot{r}_{p,x} \\ \dot{r}_{p,y} \end{bmatrix} = -k_y(c_{\Omega,y} - c_{\psi,y})R_{\theta_p} \begin{bmatrix} 0 \\ 1 \end{bmatrix}. \tag{7.16}$$

Under the assumption that the domain Ω_p encapsulates the plume, the perpendicular direction of the flux plane (w.r.t. the mean wind direction) is then used to center the flux plane in the plume.

7.2.2 Experiment

In this section, we give two illustrative simulation-based examples of using this multi-sUAS method. The first example looks at exploration versus exploitation behavior of the CVT coverage control given a static map in 2D. The second looks at implementing the CVT algorithm on a more complex 3D case with turbulence and meandering plumes using MOABS/DT [15, 13].

7.2.2.1 2D static case

The first example is in the form of a static concentration map

$$f(\mathbf{x}) = \sin\frac{x_1}{2} \sin\frac{x_2}{2} \left(\exp\frac{-(x_1 - 5)^2)}{25} + \exp\frac{-(x_2 - 5)^2}{25} \right), \tag{7.17}$$

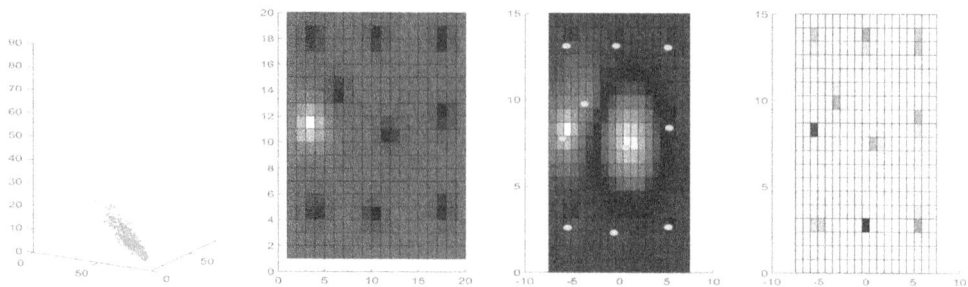

FIGURE 7.10 (left) 3D fugitive gas emission with deployed sUASs performing DE-CVT algorithm (middle-left) Ordinary kriging estimation given current concentration measurements for the sUASs at timestep k (middle-right) The first DMD mode capturing the low-frequency behavior of the kriging results and (right) the entropy map computed from the kriging estimation variance. (reused from [13] with permission from ©IEEE)

where the sUASs are deployed within a uniform ball of radius 2 located at the origin. In Fig. 7.8, we show the exploration/exploitation behavior of the density based CVT (D-CVT) and the density/entropy based CVT (DE-CVT), where the position update gains are set to: $k_p = k_e = 2$ and $k_v = 1$. The benefit of adding entropy also improves estimation error over time (see Fig. 7.9).

7.2.2.2 3D flux plane case

The 3D flux plane case represents a fugitive gas emission scenario. The scenario consists of a single source rate and $N = 10$ sUASs. The wind speed is relatively constant, such that $\bar{u} = [-1, 1]^T$. The state of the plume (or emission source), the wind field, and the sUAS sensor are computed using MOABS/DT. The 3D method is shown in Fig. 7.10, where the difference in the instantaneous emission estimate from kriging $\hat{\rho}$ and the DMD mode ψ_1 can be seen. The continuous source rate estimation performance shows good accuracy when measurements are more consistent and varies with the observations. This behavior can be seen in Fig. 7.11.

7.2.3 Conclusion

This case study introduced a new multi-sUAS method to make continuous source rate estimations using an optimal coverage control CVT based on DMD mode which is based density and kriging variance entropy. The 2D simulation shows that adding entropy can provide better exploration behavior. The 3D simulation shows that the continuous source rate estimation using DE-CVT provides a good estimate of the source rate given that the sUASs are inside the plume making good measurements.

For future work, we will not only look at exploring DE-CVT's limitations on source rate estimation (as a function of meteorological conditions, number of sUASs, the source rate, location, etc.) but will seek to compare existing methodologies (single-sUAS/sensor approaches such as mass balance [7], tracer correlation method [20, 10],

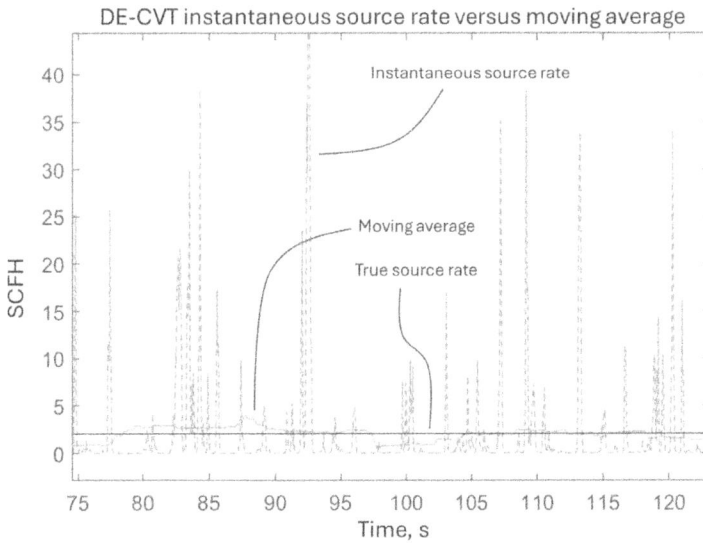

FIGURE 7.11   A snapshot of the instantaneous source rate (dotted), the continuous mass rate using 10 s moving average, and true source rate. (reused from [13] with permission from ©IEEE)

near-field Gaussian inverse [24], and OTM 33A [9]) against DE-CVT. Additionally, an evaluation metric for computing the efficiency of each method (on a per sUAS basis) will be developed. The efficiency should contain variables such as system cost, survey speed, accuracy, precision, and overall complexity. The next step is to extend to physical experiments.

7.3   EMISSION QUANTIFICATION WITH MULTI-sUAS MOD-NGI

In this section, we take a look at the mod-NGI method in the context of multi-sUAS, to see if it is possible to further improve the speed and robustness of estimating the source. To illustrate the usefulness of the mod-NGI method for multi-sUAS, a series of simulations are conducted comparing the NGI method and the mod-NGI method. The comparison will look at single-sUAS first and then finish with multi-sUAS.

7.3.1   Methods and Materials

The NGI and mod-NGI methods illustrated in this section are described in Sections 4.1.2.5 and 4.1.2.6. In this case study, it is assumed that the sUAS is sampling downwind of the source and the domain of interest (i.e. flux plane) contains the plume. The primary task is solely for quantification within the plane. It is also assumed that from the surface to 2 m is not accessible by the sUAS for safety reasons (see Fig. 7.12). Furthermore, consider that the sUAS is controllable at a high level, such that the spatial position can be represented with single or double integrator dynamics. There are two general types of path planning approaches explored here: simple fixed path planning (no vehicle dynamics) and adaptive path planning (single

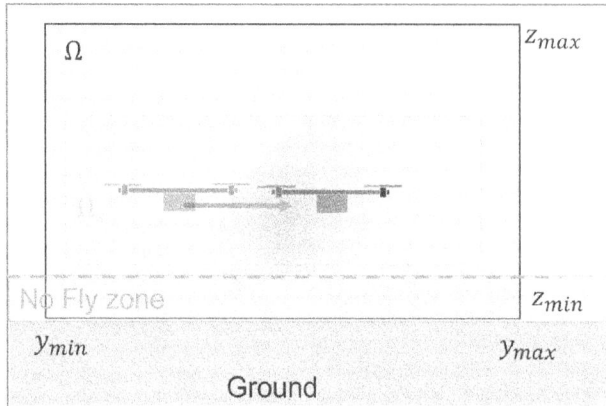

FIGURE 7.12 The simulation sample domain. (reused from [16] with permission from ©IEEE)

integrator dynamics). In the fixed path planning approaches, a lawnmower pattern (also referred to as raster scanning) and an Archimedes spiral were explored. For the adaptive path planning we look at extremum seeking control (ESC) and centroidal Voronoi tessellations (CVT) (see Fig. 7.13).

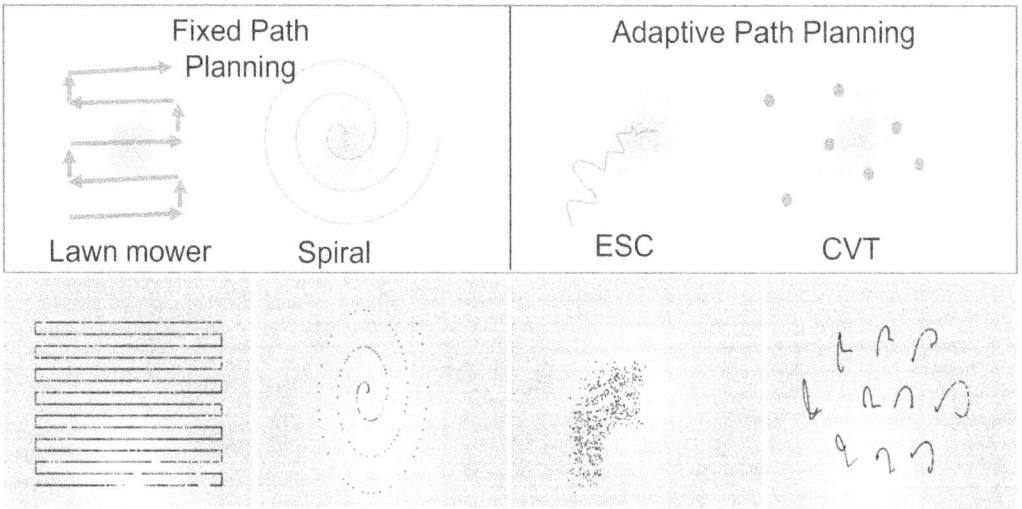

FIGURE 7.13 This image illustrates the different path planning strategies to sample the plume. (reused from [16] with permission from ©IEEE)

7.3.2 Experiment

7.3.2.1 Fixed Path Planning

To test the fixed path planning strategies a series of simulations were conducted, varying sample time, spacing between transects (or spirals), and the plume width (σ). It was observed that sample time and sample spacing (between transects) played

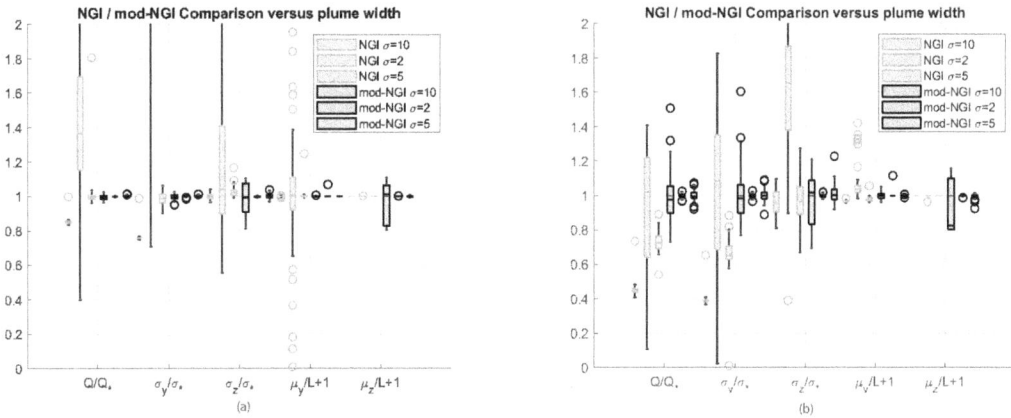

FIGURE 7.14 In these boxplots, the estimation performance results for static plumes with different plume widths are explored for (a) lawn mower and (b) spiral paths. (reused from [16] with permission from ©IEEE)

an important role in reducing the variability of the estimations. For instance, as the sample time increased (i.e. more sparse), the performance (accuracy and precision of the θ estimates) degraded. Likewise, the sample spacing reflected the ability to capture the center of the plume, often leading to poor performance. The width of the plume, in conjunction with sample time, reflected the ability to even detect the plume. Thus, the performance given different plume widths, σ, was explored for lawnmower and spiral paths (see Fig. 7.14). Conversely, it can be observed that larger plume widths will lead to better estimations and less of a dependence on which path planning approach to choose. For fugitive and small emissions this is not often the case, since at larger plume widths (further distances downwind) the concentration becomes quite small and perhaps even unmeasurable.

Given a dynamic plume setting, where the plume location has an oscillation e.g. meandering wind that can infringe on the measurement process. To test the effectiveness, the amplitude of the plume meander is increased from 0 m (no meander) to 10 m for several frequencies given the same lawnmower flight path (1 m transect spacing, 2 Hz sampling rate, 1 m/s horizontal velocity, and 10% relative noise). The results are shown in Fig. 7.15, where in some cases the source rate is not much affected and in other cases the source rate is under- and even over-estimated.

7.3.2.2 Comparison with experimental flight data

The mod-NGI method was compared with the results in [25] using their processed survey data. There were two DJI Spreading Wings S1000+ octocopter sUAS. The first was equipped with a perfluoroalkoxy tubing that tethers a local inlet on the sUAS to a ABB Micro-portable Greenhouse Gas Analyzer (MGGA) on the ground. It was also equipped with a wind sensor. The second was equipped with a lighter prototype MGGA (pMGGA). A two-dimensional stationary sonic anemometer was deployed near the boundary of the site for all surveys. The total data consisted of

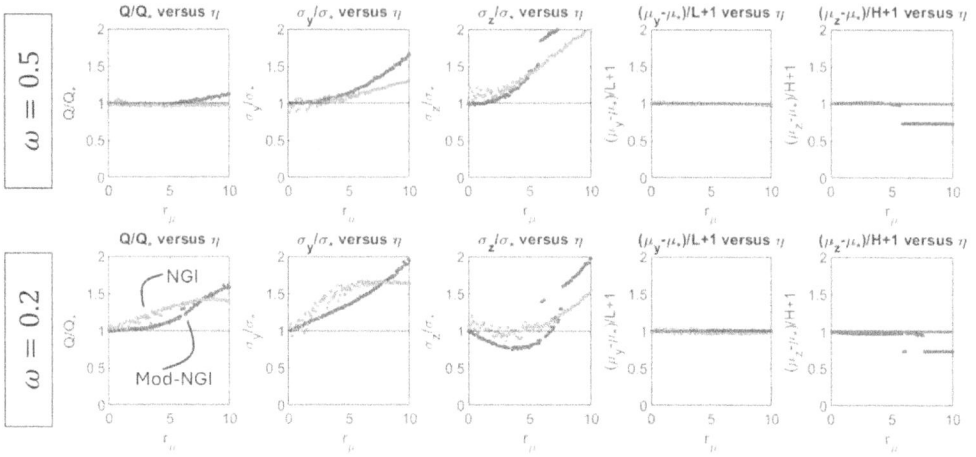

FIGURE 7.15 In these sets of plots, the performance of the NGI and the mod-NGI at different amplitudes of meandering are shown and the different frequency responses of the algorithms are shown in each row. It can be observed that both algorithms do not perform well given dynamic (e.g. meandering) conditions. The most variability can be seen in the plume width estimates. (reused from [16] with permission from ©IEEE)

22 flight surveys (7 for the first, and 15 for the second sUAS) was used to compare NGI and mod-NGI results (see Fig. 7.17). Although the mod-NGI method did not 'outperform' NGI, the results were still comparable.

7.3.3 Adaptive Path Planning

7.3.3.1 Single-sUAS extremum seeking control (ESC)

In nonlinear control theory, adaptive control is a very desirable topic, particularly if the model of the system is unknown or not accessible. Here we consider a model-free strategy called extremum seeking control (ESC), which aims to optimize a set of parameters using a perturbation signal and filtering. One of the first stability analysis works on ESC came from [19]. The ESC has been used to solve many problems, such as electric railway [21], maximum power point tracking [4], and plasma impedance matching [28]. The principle of ESC relies on the minimization of a cost function given some dynamical system. Here we treat the cost function as the concentration of the plume and seek to find the center (see Fig. 7.13). Given several values around the center of the plume, a reasonable estimate of the source could be made. However, the slope of the measurement field may not be smooth. This may lead to slow convergence times and trapping in local minima. It can be observed in Fig. 7.16d that given a dynamic plume and slow convergence, mod-NGI estimations perform poorly with ESC given the frequency of meandering.

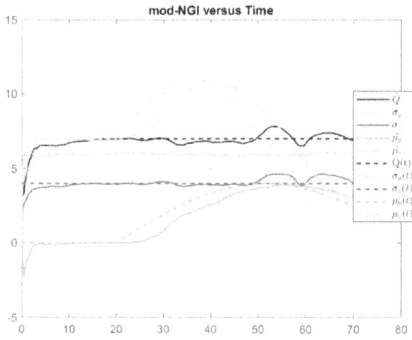

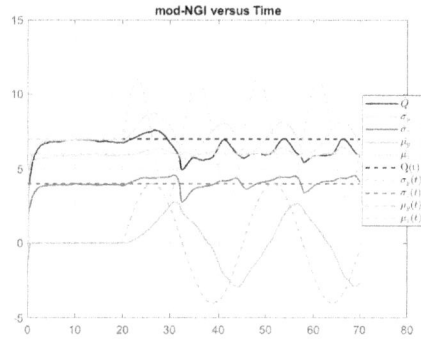

(a) CVT with $\omega = 0.1$ and 10 s window.

(b) CVT with $\omega = 0.5$ and 10 s window.

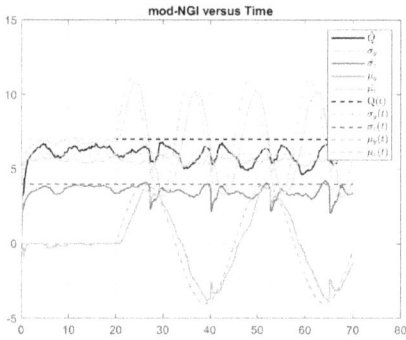

(c) CVT with $\omega = 0.5$ and 2 s window.

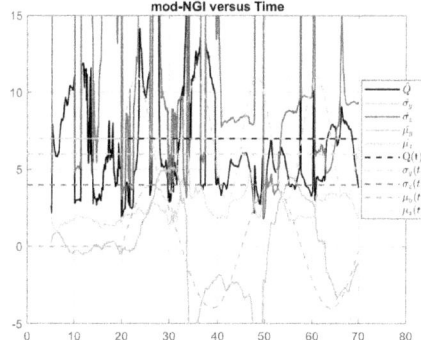

(d) ESC with $\omega = 0.5$ and 5 s window.

FIGURE 7.16 Given a moving source these six plots show the effect of the multi-sUAS CVT-mod-NGI and single ESC-mod-NGI- based algorithms. It can be observed in plots (a)–(b) that as the movement frequency of the source increases the ability of the CVT-mod-NGI algorithm to estimate and track the source diminishes. This behavior is much like the response of a time series signal given a low-pass filter. Conversely, as the window of points used to compute the mod-NGI approach is reduced, much like we would expect, the ability to track and estimate the parameters increases. Although the performance in (c) is improved, the estimations become more noisy, and thus there is a trade-off between tracking performance and stable estimations. To illustrate why multi-sUAS systems may be better, an ESC-mod-NGI algorithm is shown in (d), where the ability to estimate parameters is not as robust to meandering and exhibits more difficulty with hyperparameter tuning. (reused from [16] with permission from ©IEEE)

7.3.3.2 Multi-sUAS CVT

Taking into consideration all of the simulations thus far, they all utilize a single mobile sensor for estimation of the parameters. In this section we further explore adaptive path planning by introducing multi-sUAS coverage control strategy, namely, the centroidal Voronoi tessellations (CVT) [5] to the mod-NGI method. The CVT strategy is brought into the mod-NGI approach to improve the robustness given a dynamic

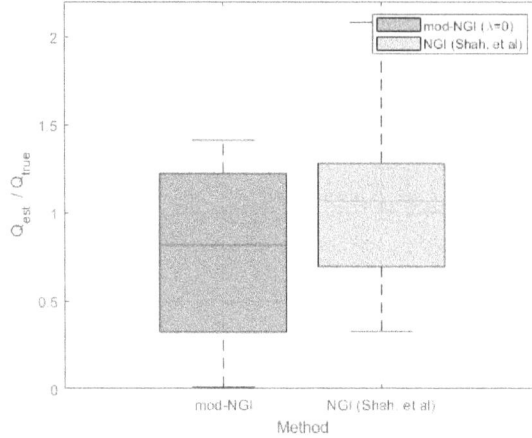

FIGURE 7.17 Comparisons between the results from Shah et al.'s NGI, our NGI, the mod-NGI, and the true controlled release flux using the controlled release survey data from [25]. (reused from [16] with permission from ©IEEE)

source (as seen in the previous sections). In this work we substitute the CVT density function for, $\rho(\mathbf{x(t)}) = \hat{y}(\mathbf{x}(t), t; \theta)$ where $\theta = [\hat{Q}, \sigma_2^c, \sigma_3^c, \mu_2^c, \mu_3^c]^T$, and the control parameters are related to the mod-NGI parameter estimations, $\theta = [\hat{Q}, \hat{\sigma}_2, \hat{\sigma}_3, \hat{\mu}_2, \hat{\mu}_3]^T$ by,

$$
\begin{aligned}
\sigma_i^c &= (1 + A_\sigma \sin(\omega_\sigma t))\hat{\sigma}_i, \quad \text{for } i = 2, 3, \\
\mu_2^c &= A_\mu \cos(\omega_\mu t) + \hat{\mu}_2, \\
\mu_3^c &= A_\mu \sin(\omega_\mu t) + \hat{\mu}_3.
\end{aligned}
\tag{7.18}
$$

The results of this approach are shown in Fig. 7.16 given $A_\mu = \omega_\mu = 1$ and $A_\sigma = \omega_\sigma = 0.5$. The control parameters are time varying to introduce a persistent excitation signal into the sampling for improving exploration and parameter estimation convergence.

7.3.4 Conclusions

In this case study we introduced the mod-NGI method for estimating emissions from mobile sensing data. The mod-NGI performance is compared with NGI in static and dynamic plume settings. The methods utilize fixed (lawnmower and spiral) and adaptive (ESC and CVT) path planning-based measurements. For single mobile sensor approaches, the proposed method shows improved estimation accuracy and precision given static plume settings, however, when exposed to a dynamic plumes the performance particularly degrades in the source rate and plume width estimations in both approaches. Regularization-based optimization was also explored and offered little improvements in certain instances and degradation in others. The mod-NGI method showed comparable estimation results to previously published NGI data using controlled release survey data. Lastly, the results of the multi-sUAS adaptive path planning approach, CVT-mod-NGI, show good tracking and estimation performance. Furthermore, the use of multi-sensor strategies, in simulation, outperform single sensor ones given a dynamic plume. This provides some motivation to utilize multi-sensor strategies, in physical experimentation. Future work will look into utilizing MOABS/DT [15] simulation (based on the ARPA-E METEC facility) to test the short time-scale behavior of the mod-NGI and CVT-mod-NGI methods. Additionally, comparison against controlled release mass balance based approaches can be found in [12].

7.4 PERFORMANCE ASSESSMENT USING MOABS/DT

In this section, we investigate how digital twins can be leveraged to conduct performance assessment between different methods. This case study leverages all of the methods learned thus far in this chapter, and some from Chapter 4. The motivations come from understanding performance limitations of a new or existing method, of which, typically require many field experiments under different scenarios and weather conditions. The methods are then validated through multiple trial runs (given each configuration), ultimately, making it costly to pursue and difficult to compare against competing methods. This is also true, going from an early method design stage to something more practical and deployable. Thus, the use of DTs in this way helps mitigate the cost of real-field testing and increases the iteration cycle of method development. In this case study MOABS/DT is used to test single and multi-sUAS emission quantification methods and their performance under different meandering and terrain conditions.

7.4.1 Experiment

To understand the performance assessment of the four quantification methods, the simulation setup was broken down into three categories: No Obstacles, Smal

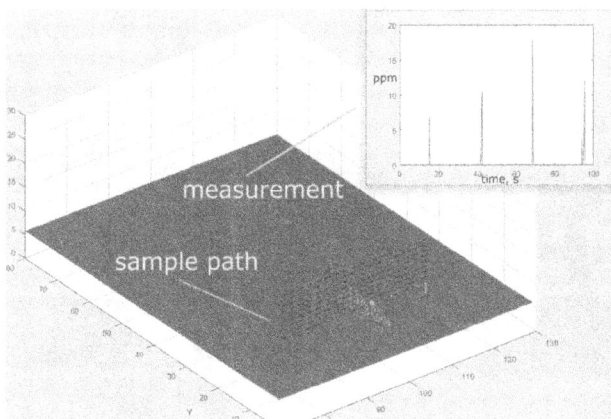

FIGURE 7.18 To validate the model the sUAS samples along a fixed path downwind of the source. The measured signal is shown in the top right. (reused from [17] with permission from ©IEEE)

TABLE 7.2   An overview of the simulation testing scenarios.

Test	Meandering		Wind Speed (m/s)	# Plumes Per Config.	Distances (From x=0)	# Plumes
Near-field and mid-field, No Obstacles	None, Large	Small, 1		5	25, 45, 75, 100, 150, 200, 275, 310, 360, 415, 500	15
Near-field, Small Obstacles	None, Large	Small, 1		5	25, 45, 75, 100, 150, 200	15
Mid-field, Small Obstacles	None, Large	Small, 1		1	275, 310, 360, 415, 500	3
Near-field, Large Obstacles	None, Large	Small, 1		5	20, 50, 70, 90, 110, 120	15

Obstacles, and Large Obstacles. The emission locations were chosen from the METEC site and surrounding area, which can be seen in Fig. 7.19. The terrain used for the METEC site is derived from LiDAR data. The terrain data is coupled with the MOABS/DT's collision algorithm, which allows for the exploration of the effects of plume-structure interaction with regard to methane source quantification. For each category, the same methane plumes were simulated, ranging from no meandering, small meandering, and large meandering. This determined how dynamic the plume was through time laterally and each of these plumes was independently simulated 5 times due to the stochasticity of the wind field. The wind conditions were chosen to be neutral with the mean wind speed near the ground being 1 m/s. The plumes were modeled with a constant small leak rate of 0.05225 g/s ($\approx$ 10 SCFH). After the plumes were fully simulated, the four methane quantification methods were tested on plumes at multiple downwind distances for each meandering type. Further details can be seen in Table 7.2 and in Fig. 7.19.

For the singe-sUAS case, the sUAS moved in a fixed path consisting of way-points constrained to a flux plane that was 50 m wide and 10 m tall. An example of this path can be seen in Fig. 7.18. The sUAS was modeled as a point mass with double integrator dynamics. The position error between the sUAS and the target position (or waypoint) was passed through a PD-controller. The sUAS was given a maximum velocity of 2 m/s and maximum acceleration of 1 m/s^2. The sUAS took 225 s to collect the methane data along its path across the fully developed plume.

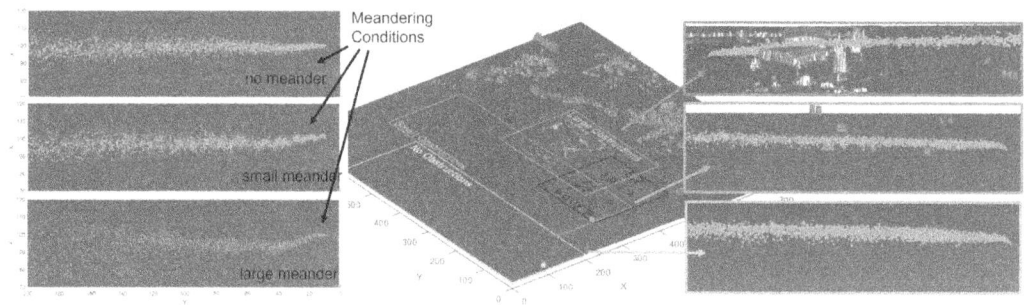

FIGURE 7.19   Subsections of the METEC location used for the case studies: No Obstacles, Small Obstacles, and Large Obstacles. The points correspond to the plume starting locations. (reused from [17] with permission from ©IEEE)

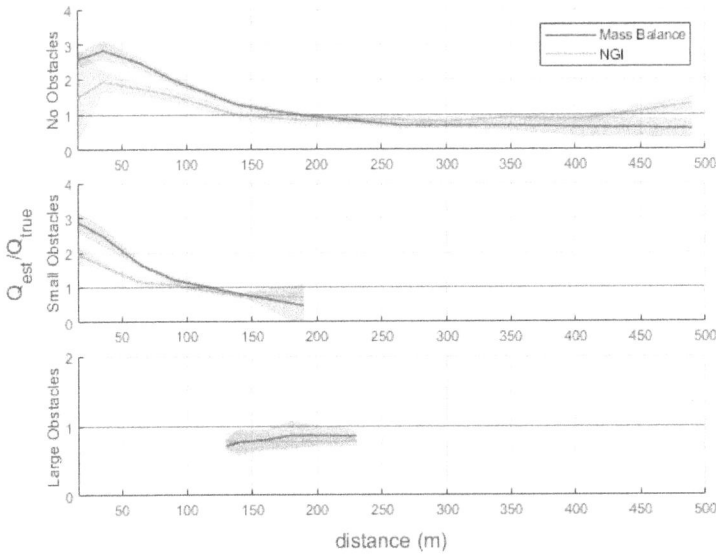

FIGURE 7.20 Results for the Mass Balance and NGI methods averaged over all of the meandering conditions and plume repetitions for each case study and distance. The solid lines indicate the means at the points, while the vertical distance from the solid line to the edge of the shaded region represents the standard deviation at that point. (reused from [17] with permission from ©IEEE)

For the multi-sUAS strategies, the swarms were given 120 seconds to sample the fully developed plume where the swarms calculated the source rate in real time. The last 96 seconds of the flux estimates were averaged to retrieve a single flux estimate for the trial.

7.4.2 Results and Observations

From the plots shown in Fig. 7.20, both of the single sUAS strategies performed well. The NGI method was observed to have performed better than the mass balance (MB) method overall, but had larger variability at 15 m from the source for the No Obstacles category. A general trend that can be seen from Figs. 7.20, 7.21, and 7.22 is that, for the No Obstacle and Small Obstacle categories, the estimates get worse closer to the source and better downwind. The Single-sUAS methods are generally best in the ranges of ≈ 100 m – 400 m. The Multi-sUAS strategies stay accurate past 400 m with the exception of DE-CVT with 3 sUASs.

When comparing all of strategies, the NGI method performed better than the MB method overall with both performing similarly in the 100 m – 400 m range. DE-CVT performed well for only a swarm size of 9 past 200 m. The Mod-NGI method performed well for both swarm sizes downwind but performed the best within 150 m of the source with a swarm size of 9. All methods displayed increased variability progressing from the no obstacles to large obstacles within the ranges tested.

The observed effect of swarm sizes in the multi-sUAS strategies indicate that as the number of sUAS increases, the accuracy of the emission estimate gets better. In

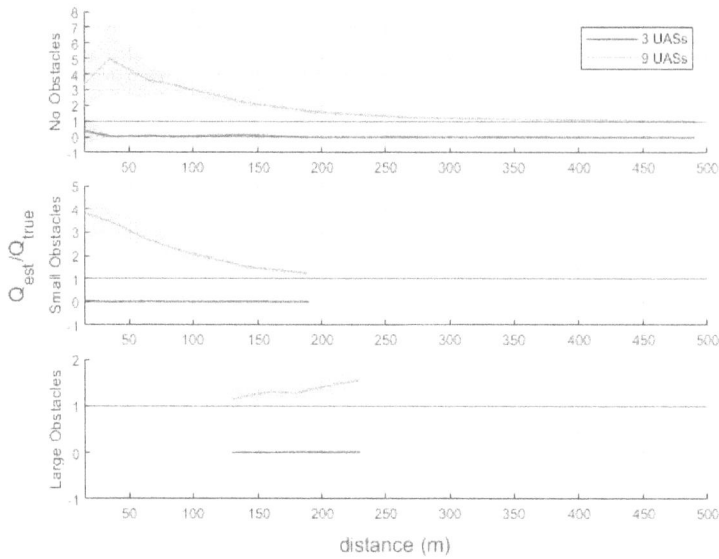

FIGURE 7.21 Results for the DE-CVT method for a swarm size of 3 UASs and 9 UASs averaged over all of the meandering conditions and plume repetitions for each case study and distance. The solid lines indicate the means at the points, while the vertical distance from the solid line to the edge of the shaded region represents the standard deviation at that point. (reused from [17] with permission from ©IEEE)

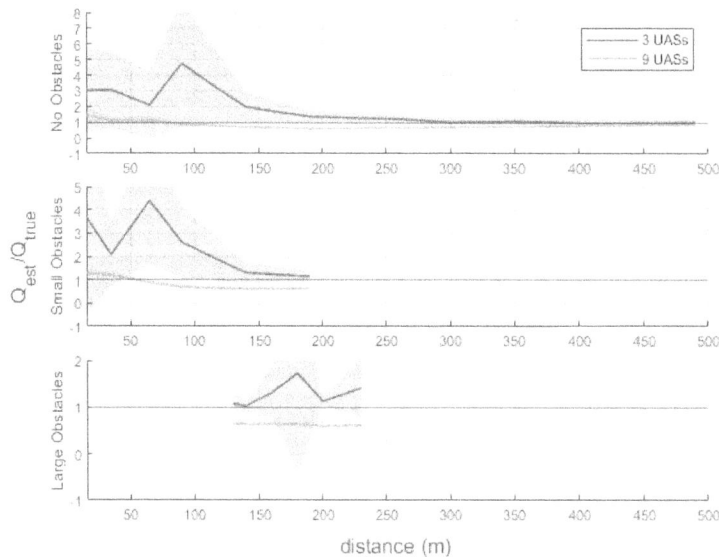

FIGURE 7.22 Results for the Mod-NGI method for a sUAS swarm size of 3 and 9 averaged over all of the meandering conditions and plume repetitions for each case study and distance. The solid lines indicate the means at the points while the vertical distance from the solid line to the edge of the shaded region represents the standard deviation at that point. (reused from [17] with permission from ©IEEE)

the DE-CVT method, the emission is marginally overestimated near the source with a swarm size of 9, whereas for a swarm size of 3, it performed poorly. In the mod-NGI case, the swarm size of 3 shows increased variability in the estimate, which indicates that there must exist an optimal swarm size for estimating the emission.

7.4.3 Conclusions

In this case study we utilize the MOABS/DT platform to assess the performance of the MB, NGI, DE-CVT, and Mod-NGI methane source quantification methods. The plume parameters were gauged via comparing methane signals in simulation to methane signals seen in practice. Results from this study show that all of the methods performed well with the single-sUAS strategies and the DE-CVT method decreasing in accuracy and increasing in variability close to the source. The mod-NGI method performance had high variability near the source for a swarm size of 3 but performed better than all of the other strategies with a swarm size of 9 at all ranges.

From this study, we can infer, as a rule of thumb, that the NGI and MB methods work best $\approx$ 100 m–400 m from the source. The DE-CVT works best for all distances beyond 200 m for higher swarm sizes. The variability of the quantification increased for all methods (given obstacle categories) except the mod-NGI method with a swarm size of nine.

An advantage of simulating these quantification methods is that it allows for the performances to be compared under the exact same environmental conditions. This eliminates any quantification variations due to changes in atmospheric conditions, allowing for a more accurate and fair means of comparison. Furthermore, utilizing DT's for benchmarking methodology is a very low-cost and efficient way to explore performance compared to field experimentation.

Based on the results observed in this study, future work will explore further the effects of swarm size on emission quantification, add more quantification methods to compare, and test strategies over various simulation scenarios.

Pause and Reflect

Now that we have seen the challenges associated with behavior matching DTs as well as the benefits of deploying DTs for method development and performance assessment, can the DT be deployed in other ways to potentially solve both problems? What are some current modeling or hardware limitations that, if solved, would enable this possibility?

7.5 CHAPTER SUMMARY

In this chapter, we presented several case studies to showcase the power of utilizing Digital Twins in the context of the emission source determination problem. The case studies presented in this chapter focused on: behavior matching raw measurement data with DT simulated data in order to estimate the DT parameters of the physical system; applying the DT model as a way to develop multi-sUAS techniques to improve emission quantification; and utilizing the DT model as a way to compare the performance between current and new methodologies. The challenges faced are primarily with the behavior matching of the physical system, as the internal model parameters of the DT require many forward runs to achieve convergence. Alternatively, utilizing new methodologies, we learned it is possible to estimate the emission rate much faster. This opens up the possibility to combine these techniques with localization techniques to constrain or provide anchor points for the behavior matching process. Future work will look into this as well as introducing atmospheric system identification using Monin-Obukhov similarity theory (MOST) [8] to improve DT model performance.

Bibliography

[1] Ricardo Andrade-Pacheco, Francois Rerolle, Jean Lemoine, Leda Hernandez, Aboulaye Meïté, Lazarus Juziwelo, Aurélien F Bibaut, Mark J van der Laan, Benjamin F Arnold, and Hugh JW Sturrock. Finding hotspots: development of an adaptive spatial sampling approach. *Scientific Reports*, 10(1):1–12, 2020.

[2] Sahar Asadi, Han Fan, Victor Hernandez Bennetts, and Achim J Lilienthal. Time-dependent gas distribution modelling. *Robotics and Autonomous Systems*, 96:157–170, 2017.

[3] Olena Babak and Clayton V Deutsch. Statistical approach to inverse distance interpolation. *Stochastic Environmental Research and Risk Assessment*, 23(5):543–553, 2009.

[4] Steven L Brunton, Clarence W Rowley, Sanjeev R Kulkarni, and Charles Clarkson. Maximum power point tracking for photovoltaic optimization using ripple-based extremum seeking control. *IEEE Transactions on Power Electronics*, 25(10):2531–2540, 2010.

[5] Jianxiong Cao, YangQuan Chen, and Changpin Li. Multi-UAV-based optimal crop-dusting of anomalously diffusing infestation of crops. In *Proc. of the 2015 American Control Conference (ACC)*, pages 1278–1283. IEEE, 2015.

[6] Lance E Christensen. Miniature tunable laser spectrometer for detection of a trace gas, June 2017. US Patent 9,671,332.

[7] OT Denmead, LA Harper, JR Freney, DWT Griffith, R Leuning, and RR Sharpe. A mass balance method for non-intrusive measurements of surface-air trace gas exchange. *Atmospheric Environment*, 32(21):3679–3688, 1998.

[8] Thomas Foken. 50 years of the Monin–Obukhov similarity theory. *Boundary-Layer Meteorology*, 119(3):431–447, 2006.

[9] Tierney A Foster-Wittig, Eben D Thoma, and John D Albertson. Estimation of point source fugitive emission rates from a single sensor time series: A conditionally-sampled Gaussian plume reconstruction. *Atmospheric Environment*, 115:101–109, 2015.

[10] Tierney A Foster-Wittig, Eben D Thoma, Roger B Green, Gary R Hater, Nathan D Swan, and Jeffrey P Chanton. Development of a mobile tracer correlation method for assessment of air emissions from landfills and other area sources. *Atmospheric Environment*, 102:323–330, 2015.

[11] Xiang He, Jake A Steiner, Joseph R Bourne, and Kam K Leang. Gaussian-based kernel for multi-agent aerial chemical-plume mapping. In *Proc. of the Dynamic Systems and Control Conference*, volume 59162, page V003T21A004. American Society of Mechanical Engineers, 2019.

[12] Derek Hollenbeck and YangQuan Chen. Characterization of ground-to-air emissions with sUAS using a digital twin framework. In *Proc. of the 2020 International Conference on Unmanned Aircraft Systems (ICUAS)*, pages 1162–1166. IEEE, 2020.

[13] Derek Hollenbeck and YangQuan Chen. Mutli-UAV method for continuous source rate estimation of fugitive gas emissions from a point source. In *Proc. of the 2021 International Conference on Unmanned Aircraft Systems (ICUAS)*. IEEE, 2021.

[14] Derek Hollenbeck and YangQuan Chen. A digital twin framework for environmental sensing with sUAS. *Journal of Intelligent & Robotic Systems*, 105(1):1, 2022.

[15] Derek Hollenbeck, Demitrius Zulevic, and YangQuan Chen. MOABS/DT: Methane Odor Abatement Simulator with Digital Twins. In *Proceedings of the 2021 IEEE 1st International Conference on Digital Twins and Parallel Intelligence (DTPI)*, pages 378–381. IEEE, 2021.

[16] Derek Hollenbeck, Demitrius Zulevic, and YangQuan Chen. A modified near-field Gaussian plume inversion method using multi-sUAS for emission quantification. In *2022 International Conference on Unmanned Aircraft Systems (ICUAS)*, pages 1620–1625. IEEE, 2022.

[17] Derek Hollenbeck, Demitrius Zulevicl, and YangQuan Chen. Single and multi-suas based emission quantification performance assessment using moabs/dt: A simulation case study. In *Proc. of the 2022 18th IEEE/ASME International Conference on Mechatronic and Embedded Systems and Applications (MESA)*, pages 1–5. IEEE, 2022.

[18] C Hunter. A recommended Pasquill-Gifford stability classification method for safety basis atmospheric dispersion modeling at SRS. Technical report, Savannah River Site (SRS), 2012.

[19] Miroslav Krstić and Hsin-Hsiung Wang. Stability of extremum seeking feedback for general nonlinear dynamic systems. *Automatica*, 36(4):595–601, 2000.

[20] Brian K Lamb, J Barry McManus, Joanne H Shorter, Charles E Kolb, Byard Mosher, Robert C Harriss, Eugene Allwine, Denise Blaha, Touche Howard,

Alex Guenther, et al. Development of atmospheric tracer methods to measure methane emissions from natural gas facilities and urban areas. *Environmental Science & Technology*, 29(6):1468–1479, 1995.

[21] Maurice Leblanc. Sur l'electrification des chemins de fer au moyen de courants alternatifs de frequence elevee. *Revue générale de l'électricité*, 12(8):275–277, 1922.

[22] Richard L Magin. Fractional calculus models of complex dynamics in biological tissues. *Computers & Mathematics with Applications*, 59(5):1586–1593, 2010.

[23] Matteo Reggente and Achim J Lilienthal. The 3D-kernel DM+V/W algorithm: Using wind information in three dimensional gas distribution modelling with a mobile robot. In *SENSORS, 2010 IEEE*, pages 999–1004. IEEE, 2010.

[24] Adil Shah, Grant Allen, Joseph R Pitt, Hugo Ricketts, Paul I Williams, Jonathan Helmore, Andrew Finlayson, Rod Robinson, Khristopher Kabbabe, Peter Hollingsworth, et al. A near-field Gaussian plume inversion flux quantification method, applied to unmanned aerial vehicle sampling. *Atmosphere*, 10(7):396, 2019.

[25] Adil Shah, Joseph R Pitt, Hugo Ricketts, J Brian Leen, Paul I Williams, Khristopher Kabbabe, Martin W Gallagher, and Grant Allen. Testing the near-field Gaussian plume inversion flux quantification technique using unmanned aerial vehicle sampling. *Atmospheric Measurement Techniques*, 13(3):1467–1484, 2020.

[26] AK Shukla and JC Prajapati. On a generalization of Mittag-Leffler function and its properties. *Journal of Mathematical Analysis and Applications*, 336(2):797–811, 2007.

[27] Gabriel A Terejanu. Discrete Kalman filter tutorial. *University at Buffalo, Department of Computer Science and Engineering, NY*, 14260, 2013.

[28] Jairo Viola, Derek Hollenbeck, Carlos Rodriguez, and YangQuan Chen. Fractional-order stochastic extremum seeking control with dithering noise for plasma impedance matching. In *Proceedings of the 2021 IEEE Conference on Control Technology and Applications (CCTA)*, pages 247–252. IEEE, 2021.

[29] Hans Wackernagel. Ordinary Kriging. In *Multivariate Geostatistics*, pages 79–88. Springer, 2003.

[30] Greg Welch, Gary Bishop, et al. An introduction to the Kalman filter. *Chapel Hill, NC, USA*, 1995.

[31] Yunfei Xu, Jongeun Choi, Sarat Dass, and Tapabrata Maiti. *Bayesian Prediction and Adaptive Sampling Algorithms for Mobile Sensor Networks: Online Environmental Field Reconstruction in Space and Time*. Springer, 2015.

CHAPTER 8

Smart Sensing, Sensor Placement, and the Observability Gramian

I N the previous seven chapters of this book, we basically learned about three parts: the methane problem, how to perform source determination with sUAS, and the basics of digital twin applications to environmental sensing. This chapter and the succeeding chapters will focus on how to use the digital twins to make the environmental sensing smarter.

8.1   WHAT IS SMART ENVIRONMENTAL SENSING?

To answer the question, we need to first reflect back on Chapter 2 and the topic of smart sensing. We discussed briefly how mobile sensing systems (e.g. sUAS) can be used to remotely gather data. With advancements in sensor technologies (e.g. sensitivities, sample frequency, lower cost, lighter weight, etc.) and advancements in IoT and edge devices, real-time optimization, and feedback are capable of occurring. When systems have this real-time feedback they inherently can become smarter and more informed about how to best sense. Introducing Digital Twins (or other advanced modeling techniques) to represent complex systems, we can begin to understand more of our environment by improving parameter estimates. For example, using the behavior matching to estimate the parameters of an emission source (or plume). This gives us key details of the emission but also allows for additional tasking, such as, utilizing the sUAS to find the optimal downwind distance or predicting the location of the plume for path planning strategies.

Smart Sensing
Is the use of advanced sensor technologies combined with data processing and communication capabilities to enhance collection, analysis, and use of the sensor data within an actionable time-scale.

We know that there are key factors from natural systems, anthropogenic systems,

DOI: 10.1201/9781003669470-8

and others (e.g. nature, mechanical wear, and the human-factor) impacting our environment and making it difficult to represent via digital twins or other surrogate models. Depending on which environmental variables we are interested in, the way we go about sensing them differs. The goal of smart environmental sensing is to utilize digital twins (or other machine learning and data-driven methods) to better inform the sensor placement and thereby generating better parameter estimations of the complex system. The National Science Foundation (NSF) has previously defined smart as having five key identifying characteristics: cognizant, taskable, reflective, ethical, and knowledge-rich. [1]

8.2 INVERSE AND ILL-POSED PROBLEMS

One of the major outstanding questions is *how to best deploy mobile sensing within the field to extract the most information?* i.e. increase the accuracy and precision of the emission estimates. Therefore, the purpose of this chapter is to explore the idea of smart sensor placement and steering for improving emission quantification. One of the major challenges of estimating emissions, (e.g. inverse dispersion methodology) using methods outlined in Chapter 4, are such that the problem is often ill-posed. Lets consider the system

$$\mathbf{u} = \mathbf{A}\mathbf{z} \equiv \int_a^b K(x,s)z(s)ds, \qquad (8.1)$$

where $K(x,s)$ is a continuous kernel with partial derivative $\partial K/\partial x$. According to Hadamard, for a problem to be well-posed it must meet three conditions [18, 19]: (1) A solution, $\mathbf{u}$, must exist; (2) There exists a unique solution, $\mathbf{u}$, for a given input $\mathbf{z}$; and (3) The inverse solution, $\mathbf{z} = \mathbf{A}^{-1}\mathbf{u}$, is stable under small perturbations (see the diagram in Fig. 8.1 for a visual representation).

For the inverse dispersion problem, inherently a part of emission quantification, this is often not the case and requires optimization and regularization to solve. Therefore, in this chapter we will focus on the sensor placement and steering problem, overview empirical Gramians, and make the connection to Digital Twins in an optimization sense.

8.3 REGRESSION TECHNIQUES

In many cases, regularization techniques such as Tikhonov and Lasso are used to solve ill-posed inverse problems, where small perturbations in the input data can lead to large variations in the solution.

Tikhonov Regularization – also known as ridge regression – is a method that stabilizes the inverse problem by adding a penalty term to the objective function. Given an ill-posed linear system $Ax = b$, the Tikhonov-regularized solution minimizes,

$$\|Ax - b\|^2 + \lambda\|x\|^2, \qquad (8.2)$$

---

[1]https://www.nsf.gov/pubs/2018/nsf18557/nsf18557.htm

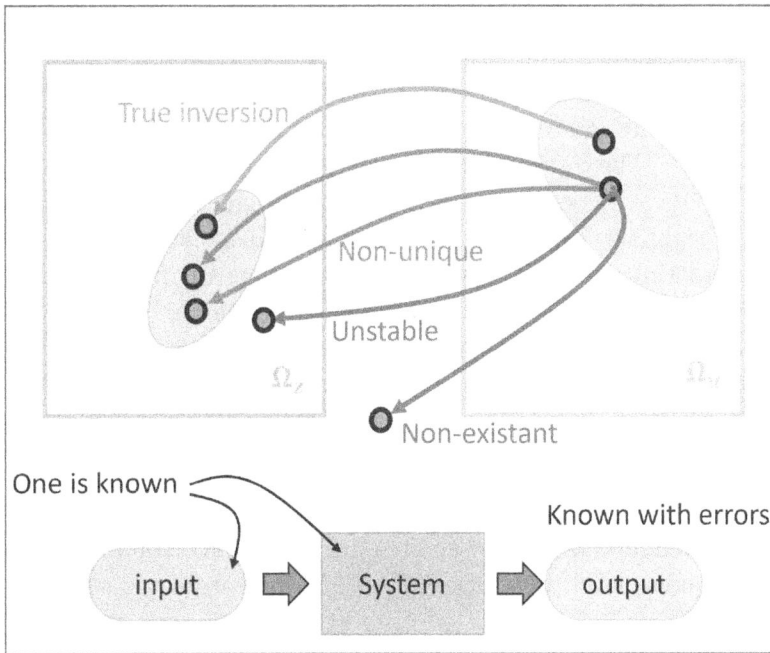

FIGURE 8.1 Inverse problem diagram.

where λ is a regularization parameter that controls the trade-off between the fidelity to the data and the smoothness of the solution. The closed-form solution is given by

$$x_{\text{Tikh}} = (A^T A + \lambda I)^{-1} A^T b. \tag{8.3}$$

Tikhonov regularization is particularly effective in suppressing high-frequency noise in the solution.

Lasso Regularization – The Least Absolute Shrinkage and Selection Operator (LASSO) – imposes an ℓ_1-norm penalty on the solution, enforcing sparsity. The objective function is

$$\|Ax - b\|^2 + \lambda \|x\|_1. \tag{8.4}$$

Unlike Tikhonov, which shrinks coefficients continuously, Lasso forces some coefficients to be exactly zero, making it useful for feature selection in inverse problems. The solution is typically obtained via iterative optimization techniques such as coordinate descent. When solving inverse problems using Tikhonov regularization, choosing the optimal regularization parameter λ is crucial for balancing data fidelity and solution stability. Two commonly used methods for selecting λ are the L-curve criterion and Generalized Cross-Validation (GCV). The interested reader should check [7] for more details in solving discrete inverse problems.

8.3.1 L-curve Criterion

The 'L-curve' is a graphical tool used to select λ by plotting the trade-off between the residual norm and the solution norm. Specifically, it plots

$$\log(\|Ax_\lambda - b\|_2) \quad \text{vs.} \quad \log(\|x_\lambda\|_2), \tag{8.5}$$

where x_λ is the solution of the regularized problem in (8.3). The plot usually forms a characteristic 'L' shape. The corner of the L-curve represents the best trade-off between minimizing the residual norm and avoiding overfitting via solution regularity. One strategy to solve for the optimal position along the L-curve is to find the point of maximum curvature. The interpretation of the L-curve plot is that the left side (i.e. small λ) corresponds to a good fit of the data, but also suffers from poor stability (e.g. overfitting). The right side of the plot (i.e. large λ) corresponds to stable solutions but suffer from underfitting data. The corner, however, offers an optimal balance between the two. The advantages of the L-curve criterion is that it is intuitive and works well for most linear problems. However, the L-curve criterion requires computing the solution for multiple λ values, and corner detection can be subjective or require automated heuristics.

8.3.2 Generalized Cross-Validation

GCV is a data-driven method that estimates the prediction error and does not require knowledge of the noise level. It is especially useful for automated selection of λ. The GCV function is

$$\text{GCV}(\lambda) = \frac{\|Ax_\lambda - b\|_2^2}{(\text{trace}(I - AA_\lambda))^2} \tag{8.6}$$

where $A_\lambda = (A^T A + \lambda^2 L^T L)^{-1} A^T$ is the regularized pseudoinverse. The regularization matrix (often $L = I$) is applied to the input vector, that is, $\lambda\|Lx\|$. For $L = I$, the GCV function simplifies to

$$\text{GCV}(\lambda) = \frac{\|Ax_\lambda - b\|_2^2}{(\text{trace}(I - A(A^T A + \lambda^2 I)^{-1} A^T))^2}. \tag{8.7}$$

The optimal λ is the one that minimizes the GCV function, thus minimizing the expected predictive error. The advantages are that the GCV can be fully automated and is suitable for problems where noise level is unknown. However, it is computationally intensive for large-scale problems due to the trace term, and can be less stable in some ill-posed scenarios.

8.4 ITERATIVE METHODS

There are several methods to solve Lasso-regularization problem. In this section we will highlight three here: iterative shrinking-threshold algorithm (ISTA), fast iterative shrinking-thresholding algorithm (FISTA) [2], and alternating direction method of multipliers (ADMM) [3].

Iterative Shrinkage-Thresholding Algorithm (ISTA) – This is a gradient-based method that iteratively updates x using

$$\mathbf{x}_{k+1} = \mathcal{T}_{\lambda t}\left(\mathbf{x}_k - 2tA^T\left(A\mathbf{x}_k - b\right)\right), \tag{8.8}$$

where t is a step-size parameter (can be set to the inverse of the Lipschitz constant, L, of $A^T A$), and $\mathcal{T}_\alpha$ is the shrinking operator,

$$\mathcal{T}_\alpha(\mathbf{x})_i = \max(|x_i| - \alpha, 0)\text{sign}(x_i). \tag{8.9}$$

ISTA converges slowly but is straightforward to implement.

Fast Iterative Shrinkage-Thresholding Algorithm (FISTA) – An improvement over ISTA with an accelerated convergence rate. It uses a momentum term to update the solution

$$x_k^* = x_k + \frac{k-1}{k+2}(x_k - x_{k-1}), \tag{8.10}$$

$$x_{k+1} = \mathcal{T}_{\lambda t}\left(x_k^* - 2tA^T(Ax_k^* - b)\right). \tag{8.11}$$

FISTA achieves a convergence rate of $O(1/k^2)$, much faster than ISTA.

Alternating Direction Method of Multipliers (ADMM) – This Splits the problem into subproblems that can be solved iteratively. ADMM introduces an auxiliary variable z such that $x = z$ and solves

$$\min_{x,z} \frac{1}{2}\|Ax - b\|^2 + \lambda\|z\|_1, \quad \text{s.t. } x = z. \tag{8.12}$$

It uses an augmented Lagrangian formulation and updates x and z iteratively, which is efficient for large-scale problems (see [3]).

TABLE 8.1 Comparison of Iterative Techniques

Method	Advantages	Limitations
ISTA	Simple to implement	Converges slowly (O(1/k))
FISTA	Faster than ISTA (O(1/k²))	More complex to implement
ADMM	Handles large-scale problems well	More computationally expensive per iteration

These iterative techniques allow solving Lasso efficiently, even for large-scale and ill-posed inverse problems. Selecting the right method depends on the problem size, sparsity structure, and computational constraints (a summary is shown in Table 8.1).

8.5 SENSOR PLACEMENT AND STEERING PROBLEM

This section aims to illustrate the sensor placement problem, which can be summarized as an optimal mobile sensor trajectory in order to achieve some desired output. In our case, the goal is to estimate the dispersion of the emission source given measured data. The sensor placement problem suggests that there is a coherent structure within the data such that the scalar field can be reconstructed with a sparse vector of measurement points. This problem is very similar to the compressed sensing problem where a transform basis $\mathbf{\Phi} \in \mathbb{R}^{n \times n}$ (typically a Fourier or Wavelet basis), and a sparse vector $\mathbf{s} \in \mathbb{R}^n$ that can reconstruct the states or signal $\mathbf{x} \in \mathbb{R}^n$,

$$\mathbf{x} = \mathbf{\Phi s}. \tag{8.13}$$

This method is typically used for image and audio signal reconstruction where they are compressible. Despite having considerable success in real-world applications it still relies on access to the full high-dimensional measurements [4]. In practice, a single mobile sensor system is deployed to make measurements of the system and the field changes spatially and temporally. Thus, the question of where best to steer the mobile sensor to achieve the best performance arises. In the work by Ucinski et al. [21], optimal sensor placement was explored for estimation of distributed parameter systems. Given a possibly nonlinear system

$$\frac{\partial y}{\partial t} = \mathcal{A}(x, t, y) + f(x, t), \tag{8.14}$$

where $\mathcal{A}$ is a spatial differential operator and f is the source term. The authors utilize time intervals (called stages) where they compute the parameters via maximum likelihood estimator. The forward problem is then computed (using numerical methods and the assumption that the wind field is known). Given the forward problem, the optimal trajectory can be computed given the Fisher Information Matrix (FIM), F, and D-optimality condition, $\Psi[F] = -\log \det(F)$,

$$F = \sum_{j=1}^{N} \int_0^{t_f} g(x^j(t), t) g^T(x^j(t), t) dt, \tag{8.15}$$

where

$$g(x, t) = \nabla_\theta y(x, t; \theta)|_{\theta=\theta^0}. \tag{8.16}$$

The concept is that the D-optimality condition-based trajectory will lead to better maximum likelihood estimates of the system parameters and thus reduce overall error. Other works have looked into this FIM way of thinking before, such as with [20], or even exploiting the gradient of the Cramer Rao lower bound in [15].

In essence, the optimal trajectory depends on what the optimality metric or condition is. For instance, there are several other metrics using the FIM. In particular, there is the E-optimality $\Psi[F] = \lambda_{max}(F^{-1})$ (or smallest eigenvalue), A-optimality $\Psi[F] = \text{tr}(F^{-1})$, or the sensitivity criterion $\Psi[F] = -\text{tr}(F)$. Other types of metrics have been explored in an effort to find better performance. For example, entropy has been used as a coverage control metric for deployment with the continuous source rate estimation method [11] described in Chapter 4, or in source seeking context with the Entrotaxis method [12]. Instead of looking for the observability of a system, there is another promising deployment metric utilizing control-theoretic technique [11]. An example in the literature by Hinson et al. [9], where they used the empirical observability Gramian as a cost function to select the best sensor placement for gyroscopic sensing in a hawkmoth wing. The results reflected quite well to what was seen in nature.

However, in order to update the sensor position or decide on a trajectory the forward model needs to be computable, at least on a timescale that is relevant to the application. To compute (8.16), the output needs to be generated for the derivative with respect to each parameter. For high dimensional systems, this can be challenging.

Another method for solving such an issue is the simultaneous perturbation and stochastic approximation (SPSA) algorithm [17]. When given a p-dimensional system, which requires $2p$ noisy measurements using finite differences with traditional stochastic approximation algorithms, the SPSA algorithm only requires $2q$, such that

$q \ll p$. There have been some advancements using digital-twin-enabled smart control engineering with parallel SPSA [22].

8.6 EMPIRICAL GRAMIANS

In control theory, the observability Gramian is an important concept for designing state estimators or performing model order reduction. However, this is typically limited to linear systems theory and not applicable to nonlinear systems. An extension to this concept is the empirical Gramian, which was developed for the use with nonlinear and parametric systems. Given the nonlinear system,

$$\dot{x} = f(t, x(t), u(t), \theta), \quad y(t) = h(t, x(t), u(t), \theta), \tag{8.17}$$

the empirical observability Gramian can be defined as

$$\hat{W}_O = \frac{1}{|S_x|} \sum_{l=1}^{|S_x|} \frac{1}{d_l^2} \int_0^\infty \Psi^l(t) dt, \tag{8.18}$$

where $\Psi^l = (y^{li}(t) - \bar{y}^{li})^T (y^{li}(t) - \bar{y}^{li})$. The initial state configuration is given as, $x_0^{li} = d_l \epsilon^i + \bar{x}$, $u(t) = \bar{u}$, and $\bar{y} = 1/T \int_0^T y(t) dt$. The term $S_x = \{d_l \in \mathbb{R} : l = 1...L, d_l \neq 0\}$. There are existing tools in MATLAB to compute W_o, such as *emgr* function (outlined in [8]).

If the system has process noise the trajectories can be computed using the forward Euler-Maruyama method for the Itô calculus, such that,

$$X_{t+1} = X_t + f(X_t, u_t)\Delta t + \sigma(X_t, u_t)Z_t\sqrt{\Delta t}, \tag{8.19}$$

where $Z_t \sim \mathcal{N}(0, I)$ and X_0 is sampled from the initial condition distribution [16]. The expectation for the empirical observability Gramian can be computed using Theorem 7 in [16],

$$E[W_O^\epsilon(t_1, x_0, u)] = \bar{W}_O^\epsilon + \hat{W}_O^\epsilon, \tag{8.20}$$

$$(\bar{W}_O^\epsilon)_{ij} = \frac{1}{4\epsilon^2} \int_0^{t_1} (E[y^{+i}] - E[y^{-i}])^T (E[y^{+j}] - E[y^{-j}]) dt, \tag{8.21}$$

$$(\hat{W}_O^\epsilon)_{ii} = \frac{1}{4\epsilon^2} \int_0^{t_1} tr(Cov[y^{+i}] - Cov[y^{-i}]) dt. \tag{8.22}$$

The stochastic empirical Gramian shows a connection with the condition number of the Fisher information matrix, $F(t)$, and the covariance matrix, R, of the measurement system [16] (see Fig. 8.2). This relation is shown as

$$F(t) \preccurlyeq \bar{\sigma}(R^{-1}) \frac{d}{dt} \lim_{\epsilon \to 0} W_O^\epsilon(t, x_0, u), \tag{8.23}$$

where $\bar{\sigma}(R^{-1})$ represents the maximum singular value of R^{-1}.

Once the empirical observability Gramian is obtained, the optimization can be done. Optimizations such as those with the objective functions, $J(\cdot)$ using the measures of unobservability [13], or with the $-tr(W^{-1})$, $\log \det(W)$, and $rank(W)$ [10].

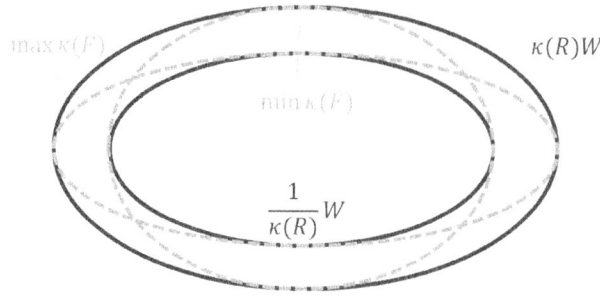

FIGURE 8.2 The minimum and maximum condition numbers of the Fisher information matrix is bounded by the empirical observability Gramian scaled with the condition number of the measurement noise covariance matrix (see [16] for more details).

TABLE 8.2 Observability metrics that can be used as a minimizing cost function for evaluating a system.

Metric	Significance
$\kappa(W)$	shape of estimation uncertainty ellipsoid [13]
$\sigma_{min}(W)^{-1}$	unobservability index [13]
$tr(W^{-1})$	minimize average estimation uncertainty [5]
$-\log(\det(W))$	minimize estimation uncertainty volume [5]
$-rank(W)$	max observable states [5]
$\lambda_{min}(W)^{-1}$	maximum estimation uncertainty [10]
$\lambda_{max}(W)^{-1}$	minimum estimation uncertainty [10]
$\det(W^{-1})$	volume of estimation uncertainty [10]

Because the observability Gramian is closely related to the estimation uncertainty and the FIM, the eigenvalues of the observability Gramian directly control the Fisher information and inversely control the estimation covariance. The minimum eigenvalue, $\lambda_{min}(W)$, is a measure of the output energy for the least observable mode, while the maximum eigenvalue, $\lambda_{max}(W)$, is the measure of the output energy for the most observable mode. The potential cost functions and their significance are listed in Table 8.2.

8.7 PERSISTENCE OF EXCITATION AND SUFFICIENT RICHNESS

The term, persistence of excitation (PE), commonly refers to an exogenous signal that excites the states of the system. This is important to consider in mobile sensor actuator systems as sensor position is reflected in the output only, and is needed to observe the excited states, where they are observable. The formal definition for PE is given here [1],

$$\sum_{t=t0+1}^{t0+N} x(t)x(t)^T \geq \alpha I, \tag{8.24}$$

where $x(t) \in R^N$ is a sequence and $\alpha > 0$. The term, sufficient richness (SR), is used to describe the PE of a control signal in the system that is order m, given N steps.

$$\sum_{t=t0+1}^{t0+N} \begin{bmatrix} x(t+1) \\ x(t+2) \\ \vdots \\ x(t+m) \end{bmatrix} [x(t+1)^T, x(t+2)^T, \cdots, x(t+m)^T] \geq \alpha I. \tag{8.25}$$

In the work by [14], the definition is given in integral form,

$$\int_{t_0} x(t)x^H(t)dt \geq \alpha I, \quad \forall t_0 \in R. \tag{8.26}$$

The definition for persistently exciting of order l of x is if $\underline{x}_l$ is persistently exciting,

$$\underline{x}_l = [x^{(0)T}, \cdots, x^{(l-1)T}]^T. \tag{8.27}$$

The importance of defining PE and SR can be shown in works like [6], such that, even a single sensor with motion can induce the PE condition and ensure the convergence of the spatial field estimation. Therefore, given multiple mobile sensors, the PE condition requirement should still hold, even though the spatial fields in [6] are linearly expandable with respect to the parameter to be estimated.

Pause and Reflect

What are some other systems that may benefit from the concept of smart sensing? Could optimal sensor placement be adapted for other emission source determination problems, such as continuous emission monitoring? How would it impact this sector?

8.8 CHAPTER SUMMARY

In this chapter, we introduced the concept of smart sensing as well as some important numerics to consider when implementing smart sensing. The fundamental issues of the inverse problem and how we can utilize regression to help solve them are key in many applications. However, it is important to note that these techniques are still difficult to use in near real-time applications due to the high system dimensionalities. The sensor placement and steering problem was introduced in the context of the compressed sensing problem and optimal sensing frameworks. A connection between the Fisher information matrix and the empirical Gramian was illustrated which gives a pathway to observability without many forward model computations. Lastly, the importance of persistence of excitation in estimating parameters are fundamental to this problem and need to be considered when designing experiments or applying inverse problem methodologies.

Bibliography

[1] Er-Wei Bai and Sosale Shankara Sastry. Persistency of excitation, sufficient richness and parameter convergence in discrete time adaptive control. *Systems & Control Letters*, 6(3):153–163, 1985.

[2] Amir Beck and Marc Teboulle. A fast iterative shrinkage-thresholding algorithm for linear inverse problems. *SIAM Journal on Imaging Sciences*, 2(1):183–202, 2009.

[3] Stephen Boyd, Neal Parikh, Eric Chu, Borja Peleato, Jonathan Eckstein, et al. Distributed optimization and statistical learning via the alternating direction method of multipliers. *Foundations and Trends® in Machine learning*, 3(1):1–122, 2011.

[4] Steven L Brunton and J Nathan Kutz. *Data-Driven Science and Engineering: Machine Learning, Dynamical Systems, and Control.* Cambridge University Press, 2019.

[5] Fabrizio L Cortesi, Tyler H Summers, and John Lygeros. Submodularity of energy related controllability metrics. In *53rd IEEE Conference on Decision and Control*, pages 2883–2888. IEEE, 2014.

[6] Michael A Demetriou. Inducing persistence of excitation through sensor motion in the adaptive estimation of spatial fields. In *2022 American Control Conference (ACC)*, pages 1673–1678. IEEE, 2022.

[7] Per Christian Hansen. *Discrete Inverse Problems: Insight and Algorithms.* SIAM, 2010.

[8] Christian Himpe. emgr—the empirical Gramian framework. *Algorithms*, 11(7):91, 2018.

[9] Brian T Hinson and Kristi A Morgansen. Gyroscopic sensing in the wings of the hawkmoth manduca sexta: the role of sensor location and directional sensitivity. *Bioinspiration & Biomimetics*, 10(5):056013, 2015.

[10] Brian Thomas Hinson. *Observability-Based Guidance and Sensor Placement.* PhD thesis, 2014.

[11] Derek Hollenbeck and YangQuan Chen. Mutli-UAV method for continuous source rate estimation of fugitive gas emissions from a point source. In *Proc. of the 2021 International Conference on Unmanned Aircraft Systems (ICUAS)*. IEEE, 2021.

[12] Michael Hutchinson, Hyondong Oh, and Wen-Hua Chen. Entrotaxis as a strategy for autonomous search and source reconstruction in turbulent conditions. *Information Fusion*, 42:179–189, 2018.

[13] Arthur J Krener and Kayo Ide. Measures of unobservability. In *Proceedings of the 48h IEEE Conference on Decision and Control (CDC) held jointly with 2009 28th Chinese Control Conference*, pages 6401–6406. IEEE, 2009.

[14] Niklas Nordström and Sosale Shankara Sastry. Persistency of excitation in possibly unstable continuous time systems and parameter convergence in adaptive identification. In *Adaptive Systems in Control and Signal Processing 1986*, pages 347–352. Elsevier, 1987.

[15] Boaz Porat and Arye Nehorai. Localizing vapor-emitting sources by moving sensors. *IEEE Transactions on Signal Processing*, 44(4):1018–1021, 1996.

[16] Nathan Powel and Kristi A Morgansen. Empirical observability gramian for stochastic observability of nonlinear systems. *arXiv preprint arXiv:2006.07451*, 2020.

[17] James C Spall. Multivariate stochastic approximation using a simultaneous perturbation gradient approximation. *IEEE Transactions on Automatic Control*, 37(3):332–341, 1992.

[18] AN Tikhonov and VY Arsenin. *Solutions of Ill-Posed Problems*. V.H. Winston & Sons, 1977.

[19] Andrei Nikolaevich Tikhonov, AV Goncharsky, VV Stepanov, and Anatoly G Yagola. *Numerical Methods for the Solution of Ill-Posed Problems*, volume 328. Springer Science & Business Media, 1995.

[20] Christophe Tricaud and YangQuan Chen. *Optimal Mobile Sensing and Actuation Policies in Cyber-Physical Systems*. Springer Science & Business Media, 2011.

[21] Dariusz Ucinski and Maciej Patan. Sensor network design for the estimation of spatially distributed processes. *International Journal of Applied Mathematics and Computer Science*, 20(3):459, 2010.

[22] Jairo Viola and YangQuan Chen. *Digital-Twin-Enabled Smart Control Engineering: A Framework and Case Studies*. Springer Nature, 2023.

CHAPTER 9

Case Studies: Smart Sensing

U P until now, we have discussed digital twins in the context of representing an emission source and using the digital twin as a way to develop new methods for quantifying methane emissions, as well as to conduct performance assessments on those methods. This was motivated by why methane is important and the difficulties in conducting real world tests to develop methods to detect/locate/quantify an emission source with sUAS. In Chapter 8 we learned about smart sensing and the difficulties around optimal sensor placement. However, we also learned about the benefits of sensor placement and how empirical Gramians can be used to help leverage the fact. In this chapter, we will learn about how digital twins can be used to gain information from the physical system through behavior matching and behavior forecasting.

9.1 BEHAVIOR MATCHING AND LOCALIZATION USING FIXED SENSORS

In this section we will take a look at a continuous emission monitoring scenario where a distribution of sensors may be placed across an area of interest to detect and localize an emission source rapidly. The challenge is determining the accuracy and speed to which the emission can be localized.

9.1.1 Methods and Materials

Suppose that within an environment of domain Ω, an unknown point source gas release exists at (x_s, y_s). Other than doing manual surveys (e.g. mobile sensing), the alternative solution is to use fixed sensors, spread throughout the landscape. The placement of these sensors would then depend on the number of sensors considered and where the emission source is likely to be. For a uniform landscape with uniform likelihood of the source position and no obstructions, the sensors take the form of a grid. The spacing of the grid would then depend on the number of sensors (see Fig. 9.1). The goal of this method is to determine the source location based on sparse sensor measurements.

DOI: 10.1201/9781003669470-9

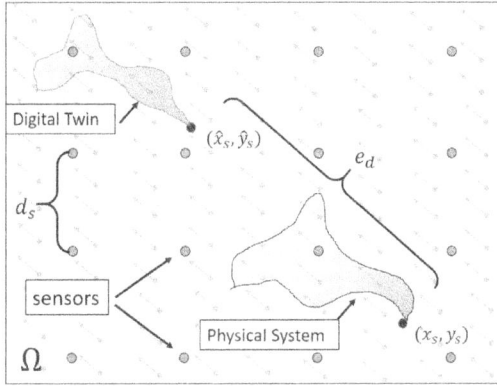

FIGURE 9.1 Given an unknown source location (x_s, y_s) the DT is behavior matched to minimize the error e_d in the source location estimation $(\hat{x}_s, \hat{y}_s)$, given a sensor distance described by d_s. (reused from [13] with permission from ©ASME)

By creating a DT of the system we can analyze the physical and simulated measurements to estimate parameters such as source rate $\hat{q}$ and source location $(\hat{x}_s, \hat{y}_s)$. However, the DT needs to first be behavior matched. Let the DT and physical system be modeled using POSIM [8, 11]. For simplicity, let's consider only the source location problem given N sensors fixed in space. At time k, the i-th sensor location is at $\mathbf{x}_i$ and measures $\rho_{i,k}$. Since the measurements are sparse we can utilize spatial interpolation (e.g. IDW) to improve some of the resolution, from N point source measurement vector ρ_k to a $n \times m$ matrix $\hat{\rho}_k \in \Omega$. Since the wind field is changing and the plume can meander, the observed measurements tend to be noisy. In order to perform behavior matching we need the low frequency behavior of the system to be slower than the control input. In this case, the control is the source location estimate $(\hat{x}_s, \hat{y}_s)$. Utilizing DMD, the low frequency modes can be retrieved from $M + 1$ snapshots of the measurement data directly,

$$Y_1 = \left[\begin{array}{cccc} | & | & & | \\ \hat{\rho}_1 & \hat{\rho}_2 & \cdots & \hat{\rho}_M \\ | & | & & | \end{array} \right], \quad Y_2 = \left[\begin{array}{cccc} | & | & & | \\ \hat{\rho}_2 & \hat{\rho}_3 & \cdots & \hat{\rho}_{M+1} \\ | & | & & | \end{array} \right], \tag{9.1}$$

$$\Psi = \mathbf{Y}_2 \mathbf{V}_Y \Sigma_Y^{-1} \mathbf{W}_Y = \left[\begin{array}{cccc} | & | & & | \\ \psi_1 & \psi_2 & \cdots & \psi_r \\ | & | & & | \end{array} \right]. \tag{9.2}$$

The recovered reduced-order DMD modes Ψ show the low frequency behavior of the spatially interpolated estimates $\hat{\rho}$ (see Fig. 9.2) such that the r-th mode ψ_r moves slow compared to the control. By performing DMD on both the physical system and the DT, the source location distance error can be defined as

$$e_d = |\mathbf{x}_s - \hat{\mathbf{x}}_s|, \tag{9.3}$$

$$\mathbf{x}_s = \arg\max_{\mathbf{x} \in \Omega} |\psi_r|, \quad \hat{\mathbf{x}}_s = \arg\max_{\mathbf{x} \in \Omega} |\psi_r^{DT}|, \tag{9.4}$$

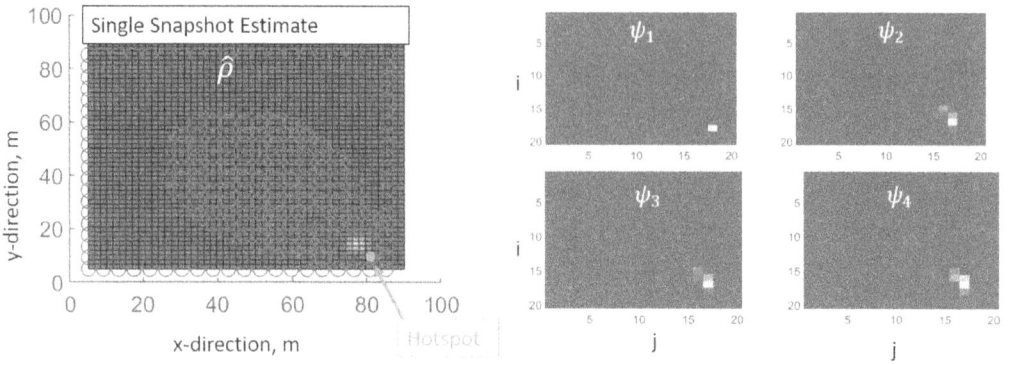

FIGURE 9.2 The results from a single snapshot of data using IDW is shown on the left and the first four spatially interpolated DMD modes are shown to the right (starting with the top row). The low frequency behavior creates a hotspot that varies less with time, allowing for behavior matching optimization. (reused from [13] with permission from ©ASME)

where ψ_r^{DT} is the DT's r-th DMD mode. The DMD modes can be selected depending on the frequency of behavior of interest. For this case the first mode is used. Given the error, a controller can be designed to provide self optimizing control (such as extremum seeking control [12, 17]) or something as simple as a proportional control law. In this case we chose the latter,

$$\dot{\hat{\mathbf{x}}}_s = k_p(\mathbf{x}_s - \hat{\mathbf{x}}_s). \tag{9.5}$$

The potential limitations to this method are with the persistent of excitation (PE) of the input signal (e.g. the plume measurements) and the physical sensor's properties. Without PE the SVD of the time snapshots can result in the eigenvalues being zero and the DMD modes cannot be computed. Additionally, the sensor's response and relaxation time can potentially limit the performance if they are too slow for the environmental conditions (or meandering wind).

9.1.2 Experiment

To illustrate the behavior matching method, we set up a simulation experiment where the source location is at $\mathbf{x}_s = (80, 10)$, and the initial guess of the DT source location to be $\hat{\mathbf{x}}_s(0) = (45, 45)$. The simulation domain spans a 90×90 unit area. The initial wind conditions are stable with no meandering ($a = 0.01$, $b = 0.01$, and $G = 1$) and unstable with meandering ($a = 0.04$, $b = 0.04$, and $G = 5$). The average wind direction is $\bar{\mathbf{u}} = [-1, 1]$. The DMD modes are computed using $M + 1 = 201$ snapshots using (50×50) IDW point estimates from 100 sparse sensors.

9.1.3 Results

The source location distance error is calculated and used to update the DT source location estimate in (9.5). The evolution of this can be seen in Fig. 9.3. The results

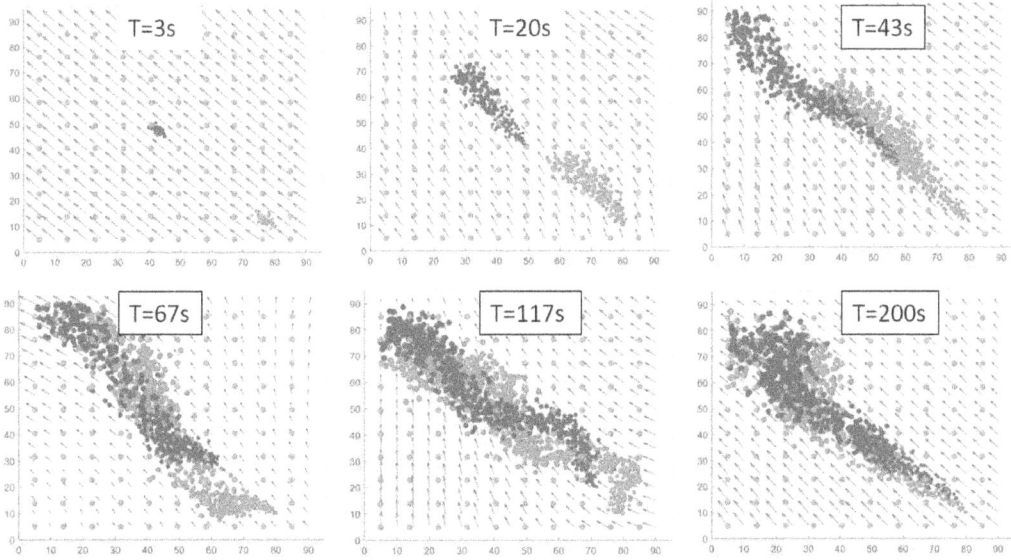

FIGURE 9.3   The time evolution of the DT behavior matching is shown for various time steps. The DT is initialized at (45,45) and the unknown physical system at (80,10). (reused from [13] with permission from ©ASME)

from the no meandering case and from the meandering case are shown in Fig. 9.4. It can be seen that the meandering wind, in slightly neutral/unstable atmospheric conditions, can push the plume to different locations in the space. For example, a source may be placed within the sparse grid of sensors, in such a way that certain wind directions may provide no observations to the sensors. However, if wind variability is introduced, the plume may redirect onto other nearby sensors, where observations can be made. Depending on the source location and sensor density, this can be seen as favorable; however, further investigation is required. Here we define the sensor density, ϱ_s, as the number, n_s, of equally spaced and distributed sensors across a domain area, Ω, divided by that area, A_Ω, such that $\varrho_s = n_s/A_\Omega$. Preliminary results on the impact of sensor density and meandering can be seen in Fig. 9.5.

9.2   DIGITAL TWIN-ENABLED SMART SENSOR PLACEMENT AND STEERING USING EMGR

The introduction of DTs into the sensor placement and steering problem provides a pathway for computing the forward problem efficiently and quite possibly in near real-time. One way to approach the sensor placement problem would be to follow the work in [16], where the sensor deployment would be updated via one of the Fisher information matrix-based metrics (e.g. D-optimality). While this seems very enticing to pursue, the Fisher information matrix requires the computation of the partial derivative of the output with respect to the parameters, leading to more forward model computations to achieve. In the context of DTs, this would require many active DT assets simulating perturbations within the parameter space. However, as

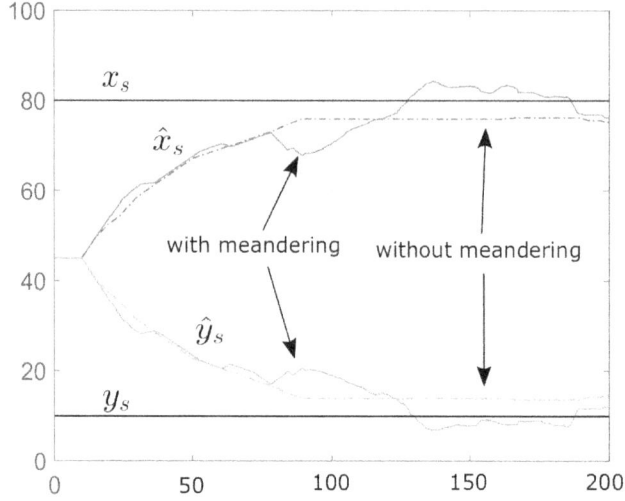

FIGURE 9.4 Given both meandering and no meandering wind conditions, the behavior matching method converges around the true source location. Even though the additional turbulence from the meandering conditions makes the convergence noisy, the estimation of the source location seems to improve. (reused from [13] with permission from ©ASME)

we seen in previous sections, the empirical observability Gramian can be computed with just the output of the system.

One of the advantages of having an active behavior matching DT of the system is the ability to converge overtime, assuming the physical system is observable and the information is sufficiently rich, the sensor placement would have need to have persistence of excitation condition to promote parameter estimation convergence [7]. To first expand on this concept, let us consider that the empirical observability Gramian can be applied to find an optimal trajectory, reducing uncertainty and improving parameter estimations.

9.2.1 Method and Materials

Let us define a domain $\Omega \subseteq \mathbb{R}^3$ that contains a single concentration source and is perturbed by an unknown wind field $\mathbf{u}$. Let there be N point measurements using mobile sensors (i.e. sUAS) deployed in Ω_p using CVT-based coverage control. The CVT plane, $\Omega_p \subseteq \Omega$, is defined by the position vector $r(t)$, the width w, the height h, and the normal vector heading φ, such that $\Omega_p \perp \underline{\Omega}$, where $\underline{\Omega}$ is the floor of Ω. The position of the i-th sensor is given as $\mathbf{z}_i(t)$ and is used to measure the concentration and wind fields. Let the concentration field be described by

$$\partial_t \rho = \nabla \cdot (D\nabla\rho) - \underbrace{\mathbf{u} \cdot (\nabla\rho)}_{\text{coupled disturbance}} + \underbrace{q\delta(\mathbf{x} - \mathbf{x}_s)}_{\text{control}}, \qquad (9.6)$$

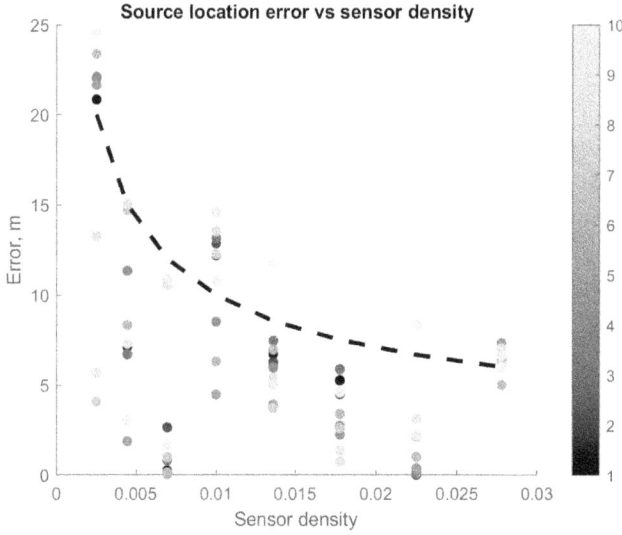

FIGURE 9.5 Preliminary results in source localization error are shown as a function of sensor density. A single source location, initial guess, and mean wind direction are used to vary meandering amplitude $1 \leq G \leq 10$ (where 10 represents larger meandering) for a fixed natural frequency $\omega_n = 0.2$ and damping ratio $\zeta = 0.1$ ($a = \omega_n^2$ and $b = 2\omega_n\zeta$). The dashed line represents the distance between adjacent sensors, d_s, and illustrates when the proposed method provides some improvements in resolution – with respect to the localization. (reused from [13] with permission from ©ASME)

where the advection component is thought of as the coupled disturbance and the source component the control signal. Then, the concentration states can be written as

$$\dot{\rho} = f(\rho, \mu, \mathbf{u}|D), \tag{9.7}$$

where $\mu = q\delta(\mathbf{x} - \mathbf{x}_s)$. Let us assume that the wind field disturbance can be described by the viscous Burgers equation,

$$\partial_t \overline{\mathbf{u}}_i = -\overline{\mathbf{u}} \cdot (\nabla \overline{\mathbf{u}}_i) + \mathbf{K} \cdot \nabla^2 \overline{\mathbf{u}}_i, \tag{9.8}$$

where $(\overline{u}_i)$ is the average wind in the i-th direction of $\mathbf{u} = [u_1, u_2, u_3]^T$, and $\mathbf{K} = [K_1, K_2, K_3]^T$ are the viscosity terms (see more details in Chapter 6).

Let's assume the mobile sensors can be modelled as single integrator systems and have negllgible effect on the physical measurement (i.e. no sensor noise). If the domain Ω is spatially discretized to support a high-dimensional number of states, the individual persistence of excitation of any $\rho_{ijk}(t)$ states can be explored. Considering the definition in (8.27), it is easy to argue that the areas with persistence of excitation are around areas with concentrations that change or are non-zero (relative to background level). This is because the control (i.e. source) is acting as the sufficiently rich signal that interacts with its neighbors according to (9.6) and (9.8). Therefore, sampling around areas, where the states are excited (given the diffusive behavior of the physical system), are of utmost importance for parameter estimation (i.e. the sensor placement problem). Thus, the problem is 1) how to choose the best place to measure and 2) how to deploy the sUAS to measure, reduce uncertainty and increase parameter estimation accuracy.

To solve the problems listed above, we have to define several smaller tasks. The first task (T1), would be to find the general area where the plume is likely to be (i.e. sufficiently rich). The approach taken here is to control the location of the domain Ω_p with $r(t)$ and φ, (referred to as a plane control task) using the empirical observability Gramian and digital twins (MOABS/DT). By controlling the domain Ω_p, the sUAS can be controlled in a stable and bounded way. It is assumed that the initial deployment of the sUAS into the plane will be undergone before the task is started. Additionally, the plane controller will actuate the plane at slower rate than the dynamics of the sUAS will allow. This is to ensure that the plane does not outpace the sUAS and potentially result in a collision. It is further assumed that the plane control will be subject to constraints, such as restricted areas (e.g. buildings or equipment) within the domain $\underline{\Omega} \subseteq \Omega$, which indicate the points on the floor of Ω, that are not restricted.

The second task (T2) would aim to solve the deployment of the sUAS inside the control plane Ω_p. In previous works, the CVT method has been used to determine the desired location of the sensors, this includes the DE-CVT and mod-NGI-CVT methods. However, the sensor locations are relatively slow moving within the domain. This slow moving (sometimes fixed) behavior may limit the sensor from reaching the states that are excited (or have concentration at some time t). Therefore, there needs to be continuous movement within the plane to increase the chances of measuring an excited state. Furthermore, in [1], the reconstruction of spatial temporal field was explored using radial basis functions (RBF). One of the major assumptions to the Theorems proposed by [1] – leading to parameter convergence – is that the sensor must pass through the center of the RBFs used to estimate the spatial-temporal field (i.e. the point of maximum concentration for Gaussian). As a result, the concept of phase portraits is used to construct a velocity field with the stable points centered inside the control plane Ω_p to induce the PE condition. The mod-NGI-CVT-based desired positions can be included in the computation to result in new objective function for obtaining a continuously moving coverage control method.

The third task (T3) is estimating the parameters of the system. This comes in three subtasks: parameter estimation, localization, and behavior matching. For parameter estimation the mod-NGI [15] algorithm proposed in task two can be used to extract the source rate and local (in plane) dispersion information. Then, a localization algorithm, such as the Lambert W function (see Section 3.2.3), particle filter, or Bayesian approach (see Section 4.1.2) can be used to estimate the source position. The estimated parameters are then mapped to the DT parameters as a part of the behavior matching process.

When all three tasks are solved iteratively, the system parameters are expected to converge to the correct values. The reasoning is that in T1, the DT is used to understand the states of the system (increasing observability for T2), then the results from T2 are used to inform T3, and ultimately, T3 improves the DT's ability to

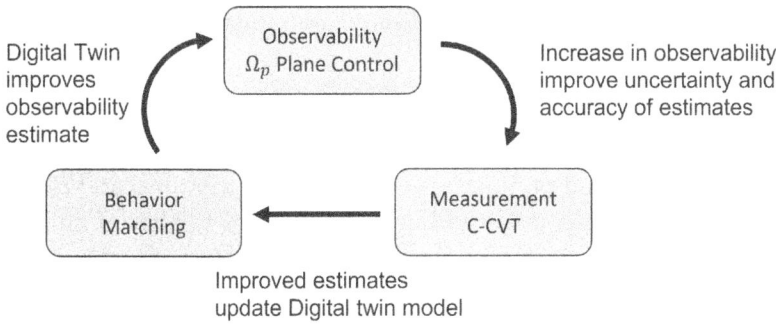

FIGURE 9.6   A diagram depicting the interconnection between the three major tasks of digital twin-enabled smart sensor placement framework.

represent the physical system (which is used in T1). The interconnection for the three tasks are shown in Fig. 9.6.

9.2.1.1   Framework

To test the performance of the DT-enabled sensor placement and steering given different empirical observability Gramian metrics, the MOABS/DT platform was explored. To better understand the performance of each of the metrics in Table 8.2, the scope of the simulation is reduced to just focus on the first task, T1. Therefore, following the framework it is assumed that the DT has been sufficiently matched with the physical system to meet the goals of tasks T2 and T3. This means that the simulation (see Fig. 9.7) includes a DT representation for the wind field and plume as well

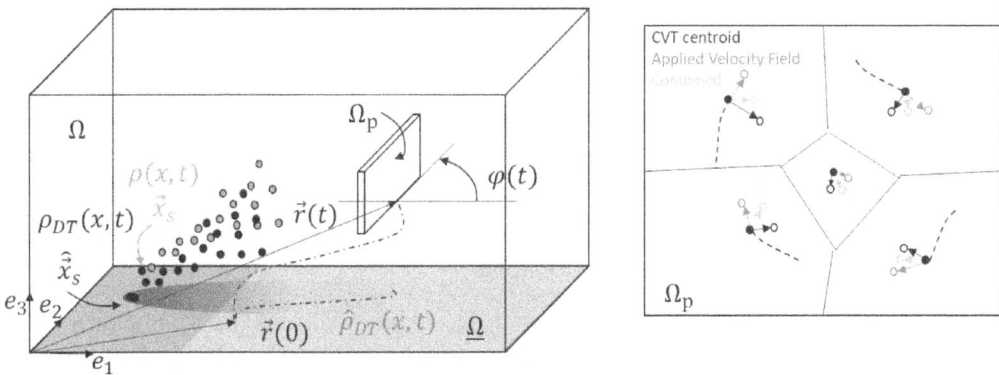

FIGURE 9.7   The domain Ω contains the source ρ, digital twin ρ_{DT}, and the Ω_p plane (functions of $r(t)$ and $\varphi(t)$). The projection from the Lagragian states to Eulerian states onto $\underline{\Omega}$ are shown by $\hat{\rho}_{DT}$. Inside $\underline{\Omega}$ there may be regions that $r(t)$ cannot go, illustrated in the shaded region near source. The CVT continuous coverage controller is shown on the right. The combination of the two desired setpoints are used to dynamically steer the centroids in an orbit around the stable point (located in the center of Ω_p).

as a simulation of the physical wind field and plume (which is used to take sensor measurements).

9.2.1.2 Ω_p Plane Controller

Given the current states and parameters of the digital twin, ρ_{DT} and θ_{DT}, a function mapping $h_{DT}(\cdot)$ is needed to map the Lagrangian states to that of an Eulerian ones $\rho_{DT} \mapsto \hat{\rho}_{DT} \in \underline{\Omega}$. This is achieved by uniformly discretizing the domain $\underline{\Omega}$ and integrating along the e_3 direction (see Fig. 9.7). The resulting map of $\hat{\rho}_{DT}$ is then used to compute the empirical observability Gramian, $\hat{W}_O$. Given the nonlinear system,

$$
\begin{aligned}
\dot{\rho}_{DT} &= f(\rho_{DT}, \mathbf{x}, \mu, \mathbf{u}, \hat{\theta}), \\
\mathbf{y} &= h_{DT}(\rho_{DT}, \mathbf{r}), \\
\dot{\mathbf{r}} &= -k_r(\mathbf{r} - \mathbf{r}_d), \\
\dot{\varphi} &= -k_\varphi \Delta \varphi_d,
\end{aligned}
\tag{9.9}
$$

where $\mathbf{r}_d$ and φ_d are the desired plane location and heading, respectively, and $k_r > 0$ and $k_\varphi > 0$ are proportional constants. To avoid computing $\hat{W}_O$ across the whole domain, only points near the current plane location are considered. Furthermore, since $\hat{\rho}_{DT} \subseteq \underline{\Omega}$ the empirical observability Gramian can be computed at some nominal point $\mathbf{r}_0$ using sample trajectories in the e_1 and e_2 directions, defined as,

$$
\hat{W}_O = \frac{1}{|S_x|} \sum_{l=1}^{|S_x|} \frac{1}{d_l^2} \int_0^\infty \Psi^l(t) dt,
\tag{9.10}
$$

where $\Psi^l = (y^{li}(t) - \bar{y}^{li})^T (y^{li}(t) - \bar{y}^{li})$. The initial state configuration is given as, $r_0^{li} = d_l e_i + \bar{r}$, $u(t) = \bar{u}$, and $\bar{y} = 1/T \int_0^T y(t) dt$. The term $S_x = \{d_l \in \mathbb{R} : l = 1...L, d_l \neq 0\}$. There are existing tools in MATLAB to compute W_o, such as *emgr* function (outlined in [9]). Given the estimate of the empirical observability Gramian in the perturbation directions, an objective function $J(W_o)$ can be chosen to evaluate each (see Table 8.2). For each direction and for each candidate objective function, the step will be evaluated and used to compute a plane control vector.

9.2.1.3 CVT Continuous Coverage Controller

Given local coordinates of the domain Ω_p, the positions of the sensors $\mathbf{z} = [\mathbf{z}_1, \mathbf{z}_2, ...\mathbf{z}_N]$ can be divided into N polytopes $\mathcal{V} = (\mathcal{V}_1, \mathcal{V}_2, ...\mathcal{V}_N)$. The polytopes are determined using centroidal Voronoi tessellations (CVT)[2]. The cost function that controls the CVT desired positions is

$$
\mathcal{H}(\mathbf{z}, t) = \sum_{i=1}^{N} \int_{\mathcal{V}_i} \mathcal{M}(q, t) |\mathbf{z}_i - q|^2 dq, \quad \text{for } q \in \Omega_p,
\tag{9.11}
$$

where $\mathcal{M}(\cdot)$ is a concentration map defined by the mod-NGI algorithm in the plane, Ω_p. The partial derivative can be used with the mass, m_i, and center of mass, c_i, to identify the critical point,

$$m_i = \int_{\mathcal{V}_i} \mathcal{M}(q,t)dq, \quad c_{m_i} = \frac{1}{m_i} \int_{\mathcal{V}_i} q\mathcal{M}(q,t)dq, \qquad (9.12)$$

$$\frac{\partial \mathcal{H}}{\partial \mathbf{z}_i} = 2m_i(\mathbf{z}_i - c_{m_i})^T. \qquad (9.13)$$

This means that $c_i = \mathbf{z}_i$ for all $i = 1, 2, ..N$ is a minimizer of the CVT. Treating the relation as a gradient, Llyod's algorithm can be used given by

$$(\dot{\mathbf{z}}_i)_c = -k_c(\mathbf{z}_i - c_{m_i}). \qquad (9.14)$$

The general algorithm to compute the locations of actuators by computing the CVT [5, 6, 4] using Lloyd's method. To make the coverage control algorithm continuous, a velocity field can be applied using the following:

$$(\dot{\mathbf{z}}_i)_v = \lambda \frac{A}{|A|} \mathbf{z}_i, \quad A = \begin{bmatrix} \beta & 1 \\ \gamma & \beta \end{bmatrix}, \qquad (9.15)$$

where β, γ, and λ are constants. When $\beta > 0$, the system is diverging, $(\mathbf{z}_i)_v \to \infty$, when $\beta < 0$, the system is converging $(\mathbf{z}_i)_v \to 0$, and when $\beta = 0$, the system is critically stable. When $\gamma < -1$ the velocity field is stretched vertically, and when $0 \leq \gamma < -1$ the velocity field is stretched horizontally. Because the system is linear, the dynamics can be easily shifted (see Fig. 9.8). Additionally, by normalizing the matrix A and having a gain λ, the resulting velocity field can be scaled to appropriate magnitudes. If the distance gets too far from the stable point, the magnitude of $(\dot{\mathbf{z}}_i)_v$ can be limited, such that,

$$(\dot{\mathbf{z}}_i)_v = v_{max} \frac{(\dot{\mathbf{z}}_i)_v}{|(\dot{\mathbf{z}}_i)_v|}, \quad \text{if } (\dot{\mathbf{z}}_i)_v > v_{max}. \qquad (9.16)$$

To combine the two forms, a time step approximation of the trajectory (in the direction of the desired velocity field) is used as a desired location with weight k_v, resulting in a new dynamic law,

$$\dot{\mathbf{z}}_i = (\dot{\mathbf{z}}_i)_c - k_v[\mathbf{z}_i - ((\dot{\mathbf{z}}_i)_v \Delta t + \mathbf{z}_i)], \quad k_v > 0. \qquad (9.17)$$

If we consider that $k = k_c = k_v$,

$$\dot{\mathbf{z}}_i = -k(\mathbf{z}_i - c_{m_i} - c_{v_i}), \quad k > 0, \qquad (9.18)$$

where $c_{v_i} = (\dot{\mathbf{z}}_i)_v \Delta t$. If we consider that the density function $\mathcal{M}(q,t)$ be a radial basis function (such as the Gaussian Plume Model used in the NGI and mod-NGI) centered at the stable point of $(\dot{\mathbf{z}})_v$ dynamics, the orbiting trajectory path would be circular as well.

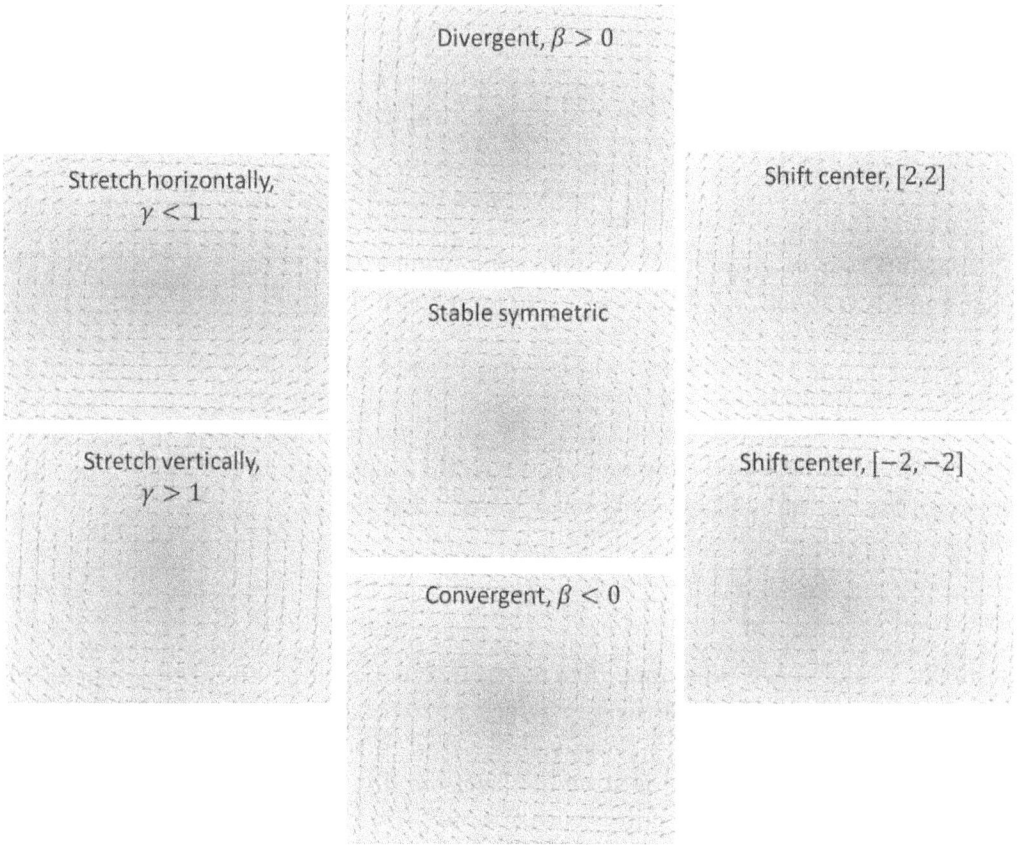

FIGURE 9.8 The phase portraits for different parameters (outlined in 9.15) can be seen here, where the velocity field is shown to be divergent/convergent, stretched, and shifted.

9.2.2 Experiment

The simulation was initialized with the true source and DT located at $\mathbf{x}_s = [120, 140, 0]$. The wind field was initialized with the mean wind vector as $\bar{\mathbf{u}} = [-1, 0]$. Utilizing the MATLAB random number generator to control the random seed, each simulation was initialized over two stages. The first stage lets the plumes develop past the Ω_p control plane as well as stabllize the sensor positions within Ω_p. Once this stage has been past, the CVT continuous coverage controller is enabled. At the second stage, the desired plane location is set to 40 meters above the current location, which is designed to cross the path of the plume 30 meters downwind (see the diagram in Fig. 9.9). When any of the sensors detects the presence of the plume (e.g. measurements significantly above background), the observability-based sensor placement and steering approach is enabled. For this experiment, the initial perturbation step size $d_l = 1$ was chosen based on the resolution of the projected DT plume $\hat{\rho}_{DT}$ onto Ω. To keep the computational cost to a minimum, we only considered a one step trajectory for each of the basis direction (e_1, e_2). For each cost function evaluation, the simulation was run for 300 seconds. During the simulation, the emission is being

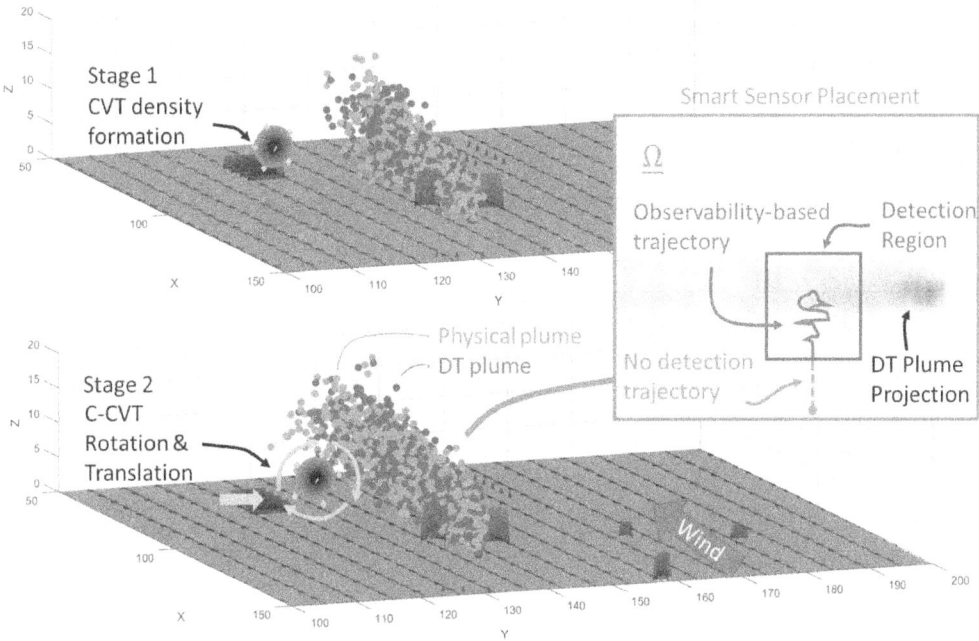

FIGURE 9.9 The MOABS/DT simulation is first initialized in (Stage 1) by letting the physical and DT plumes develop past the point of measurements, then (Stage 2) initializes the continuous CVT and commands the Ω_p plane to cross the plume. Once a detection is made, the sensor placement and steering is activated and begins using the DT projection onto $\underline{\Omega}$ for the empirical observability Gramian computation.

quantified using the mod-NGI method. In the event that the empirical observability Gramian is singular, a chemotaxis-like approach is undertaken, such that, the plane control vector $\mathbf{r}(t)$ is perturbated in the direction of the neighboring cell with the highest concentration (with respect to $\hat{\rho}_{DT}$).

9.2.3 Results

Given the first task of DT-enabled sensor placement and steering framework, the empirical observability Gramian-based cost functions were utilized to create trajectories for the Ω_p plane to estimate the emission rate. The individual trajectories are shown in Fig. 9.10 and it can be observed that several pairs of cost functions have almost identical trajectories. This is likely because each of the cost functions are trying to reduce similar forms of the estimation uncertainty and there are only two basis directions that the trajectory can take. Furthermore, there are only two pairs of trajectories that turned out to be identical, which consist of $\sigma_{min}(W)^{-1}$ and $\lambda_{min}(W)^{-1}$, and $-\log(\det(W))$ and $\det(W^{-1})$, respectively. It turns out that in this case, $\det(W)^{-1} = \det(W^{-1})$, which means they only differ from a logarithmic scaling. Furthermore, the minimum eigenvalues and minimum singular values turn out to be identical. It can also be observed that the trajectory behavior of $\sigma_{min}(W)^{-1}$,

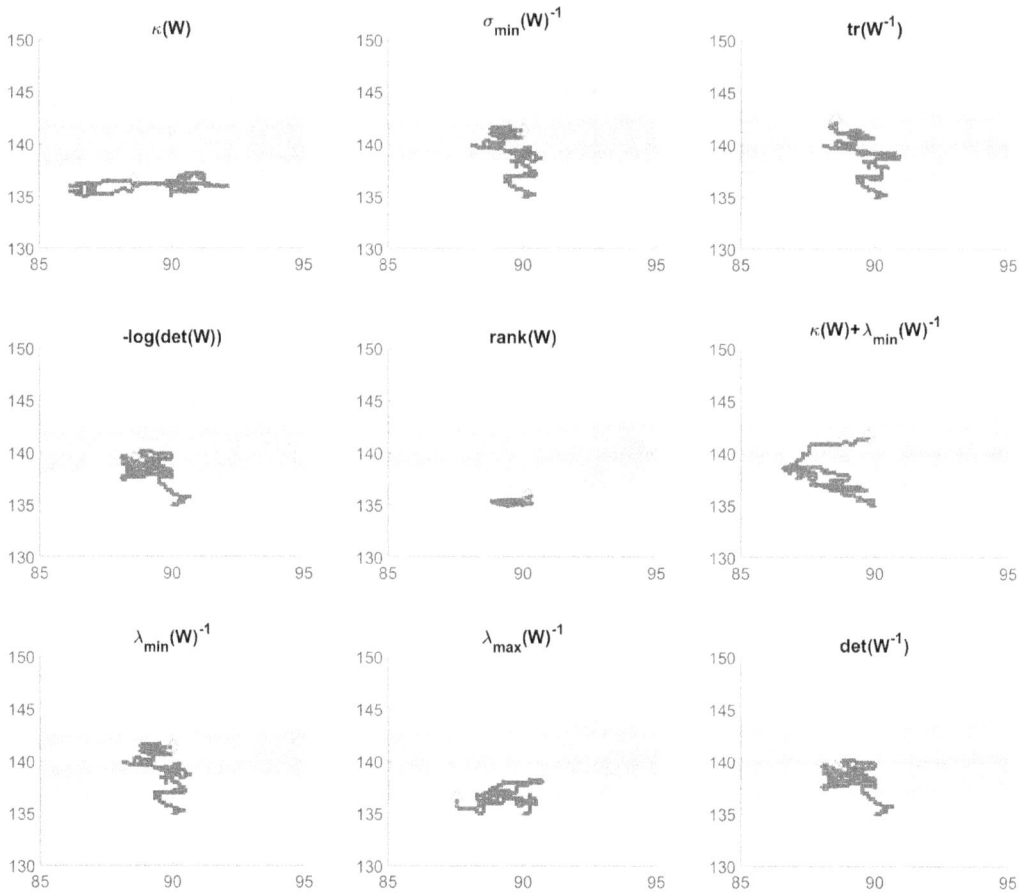

FIGURE 9.10 The sensor placement and steering trajectory for each cost function are shown here.

$\lambda_{min}(W)^{-1}$, and $tr(W^{-1})$ tend to travel in the cross wind direction, whereas, $\kappa(W)$ and $\lambda_{max}(W)^{-1}$ tend to travel in the direction of the wind. If we take a look at the linear combination of the condition number and the minimum eigenvalue inverse, $\kappa(W) + \lambda_{min}(W)^{-1}$, we can see both behaviors in the trajectory, which is unique to the weighting chosen (here it is 25). Depending on the desired behavior (given the task at hand) the weight on $\lambda_{min}(W)^{-1}$ can be increased or decreased to see an effect in the trajectory behavior. This type of steering deserves some attention and will be a topic of furture research.

Analyzing the continuous emission quantification estimates (see Fig. 9.11) shows that the cost functions with the trajectory that yielded the best results were the unobservability index $\sigma_{min}(W)^{-1}$, $\log(\det(W))$, $\lambda_{min}(W)^{-1}$, and $\det(W^{-1})$. Conversely, we see that the worst trajectory was the $rank(W)$ cost function. In fact, during the simulation if there was no general direction that would lead to an increase in the rank (i.e. the rank condition stayed the same, regardless of the trajectory direction), the sensor placement and steering strategy would reduce to chemotaxis.

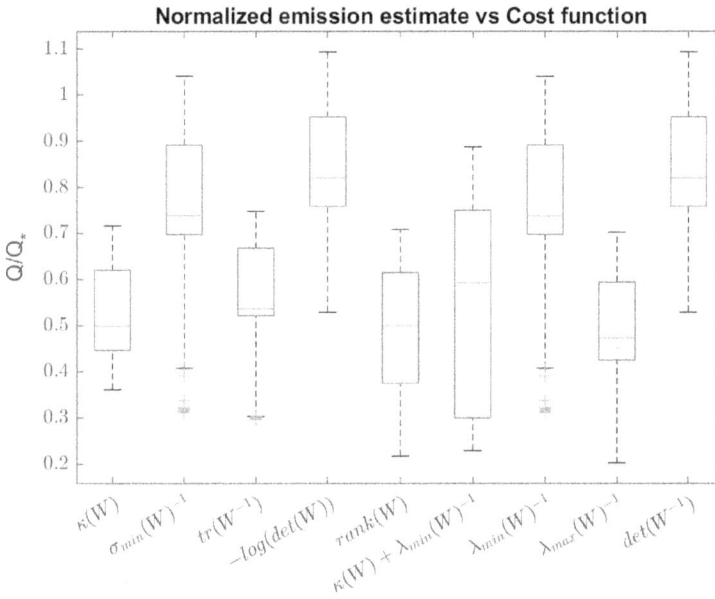

Normalized emission estimate vs Cost function

FIGURE 9.11 The emission quantification using the mod-NGI method is shown against different cost functions. The best performing metrics have smaller uncertainty and better accuracy, such as with $det(W^{-1})$.

9.3 DIGITAL TWIN-BASED ITERATIVE REFINEMENT FOR LANDFILL AREA SOURCE DETERMINATION PROBLEM

In the previous two examples, we have seen how the DT can be used to infer the location through behavior matching across multiple sensors or how the DT can be used to inform 'where to sense' through the empirical Gramians. In this section, we investigate how the DT can be used to solve the inverse problem by estimating the source location from purely trajectory data measured away from the source. This idea is useful for rapidly estimating the emission source from known methods (e.g. the mass balance method or NGI), and also gives the operators feedback on where the emission is coming from to give insights for repair or maintenance. The DT in this context provides additional details by including complex terrain data in the evolution of the plume to where it is sensed by the sUAS.

9.3.1 Methods and Materials

Given a domain of interest Ω that contains an area emission source in the subdomain $\Omega_s \subset \Omega \in \mathbb{R}^3$ with rate S, let the mobile sensor (i.e. sUAS) be actuated in Ω with a spatial position at time, t, given by, $\mathbf{x}_m(t) \in \mathbb{R}^3$, and trajectory path,

$$\mathbb{X}_m = \{\mathbf{x}_m(t)\}_{t=0}^T. \tag{9.19}$$

Given the trajectory path, let the measured signal at time, t, be defined as $\mathbf{y}_m(t) \in \mathbb{R}^1$ and the measured data set,

$$\mathbb{Y}_m = \{y_m(t)\}_{t=0}^T. \tag{9.20}$$

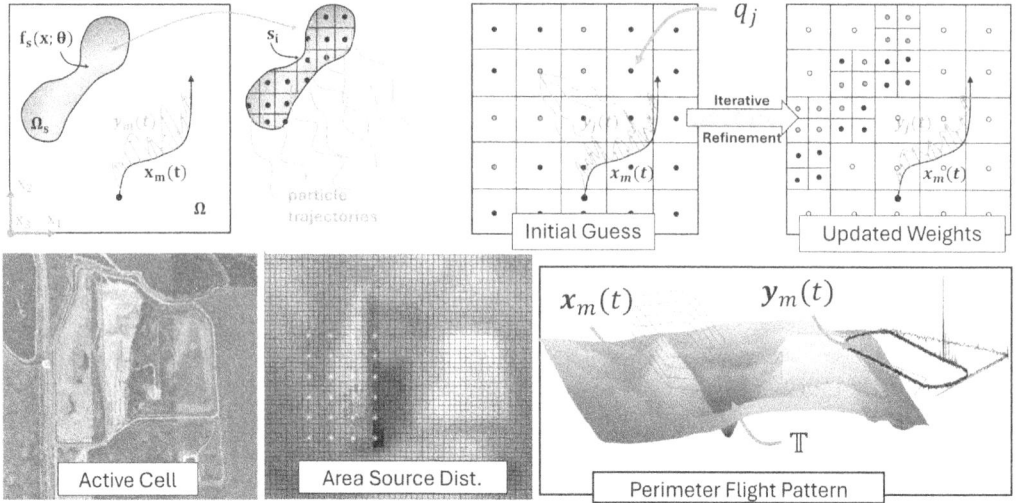

FIGURE 9.12 The coordinate system and domain for the source determination problem. The true area source is released from Ω_s, and the DT area source is released from the locations of the q_j's. Based on the DT measured signals the weights are updated. The application is centered around an active cell at the MCRWM Landfill site. The initial area source distribution (square points) is shown against the DT initial guess (asterisk points). The perimeter flight pattern is used to sample the space and generate the time series signals for the iterative refinement method.(reused from [14] with permission from ©IEEE)

Using a distribution function, $f_s(\mathbf{x}; \theta)$, the area source can be projected onto a discrete grid in Ω (see Fig. 9.12), such that the total source rate S is given as

$$S = \sum_{i=1}^{N_s} s_i \delta(\mathbf{x} - \mathbf{x}_{s_i}), \tag{9.21}$$

where $\mathbf{x} = [x_1, x_2, x_3]^T$ describes the location in Ω and $\mathbf{x}_{s_i} \in \Omega_s$ gives the spatial location of the point source emission with source rate s_i. The grid size is chosen based on the desired computational efficiency.

The spatial resolution of measurement signal will depend on the sample rate of the sensor and speed of the mobile sensor (as it transects the plume). Let the measured signal be subject to process and sensor noise, both of which are assumed to have properties of a Normal distributions $\eta(t) \sim \mathcal{N}(0, \sigma_\eta^2)$ and $\varepsilon(t) \sim \mathcal{N}(0, \sigma_\varepsilon^2)$, respectfully.

Let the trajectory path and the measured signal be related through the measurement function, given by the physical system as $h_m : \mathbf{x}_m(t) \mapsto y_m(t)$. Given that the measured signal depends on the source data $\mathbb{S} = \{(\mathbf{x}_{s_i}, s_i) | \forall i \in \Omega_s\}$, measured wind data $\mathbb{U} = \{\mathbf{u}_m(\mathbf{x}_u, t)\}_{t=0}^T$, and the knowledge about the geographic topology $\mathbb{T} = \{\mathbf{x}_g | \forall \mathbf{x} \in \Omega, s.t. x_3 = f_g(\mathbf{x})\}$, they are all required to provide a solution to the inverse problem. The function $f_g(\mathbf{x})$ represents the spatial mapping to the surface of the terrain. The measured wind signal $\mathbf{u}_m(\mathbf{x}_u, t)$ will provide input into the DT model for updating the dynamics, and thus depends heavily on the location, $\mathbf{x}_u$, of the sensor. Typically, the sensor is fixed somewhere just above the ground $(x_{u_3} > f_g(\mathbf{x}_u)$ e.g. a few meters), near the takeoff area, and away from obstructions. With the varying topology of the landfill, the choice of wind sensor location becomes more imperative, however, here we reserve the problem of optimal wind sensor placement as a topic of future research.

To solve this inverse problem, we start by proposing some simplifying assumptions: (A1) The area source quantification has already been estimated (e.g. $S = Q$); (A2) the atmospheric conditions are sufficiently measured and known (i.e. the DT plume dynamics can represent the physical system's plume dynamics); and (A3) The area source distribution estimation can be carried out in the offline sense.

Iteration Step: We begin by utilizing a probabilistic prior $p(\Theta)$ on the domain Ω, where Θ represents the set of point source locations, $\mathbb{X}_q = \{\mathbf{x}_{q_j} | j = 1, \cdots, N_q\}$ and the set of emission rates $\mathbb{Q} = \{q_j | j = 1, \cdots, N_q\}$, to be solved. Initially, we assume a uniform prior across Ω which meant that all the spatial locations were equally possible across Ω and all the emission rates were equally distributed with the assumed source rate (see Fig. 9.12),

$$q_j = Q. \tag{9.22}$$

Setting each j-th discrete emission source as the assumed source rate Q allows for the possibility of the solution of a single point source, as well as for the signal to overcome minimum measurement threshold concentrations. Depending on the size of Ω and resolution desired, the size of the grid can become smaller or larger, ultimately changing the initial number of point sources to represent the area source distribution. For each proposed point source location and rate, a forward simulation is computed using a MOABS/DT plume simulation. The resulting measurement signal data set $\mathbb{Y}_j = \{y_j(t)\}_{t=0}^T$ is gathered using the DT measurement function $h_j : \mathbf{x}_m(t) \mapsto y_j(t)$.

Let us define the likelihood function $p(\mathbb{D}|\Theta)$ that takes in the whole data $\mathbb{D} = \{\mathbb{X}_m, \mathbb{Y}, \mathbb{U}, \mathbb{T}\}$ and parameters $\Theta = \{\mathbb{X}_q, \mathbb{Q}\}$, where $\mathbb{Y} = \{\mathbb{Y}_m, \mathbb{Y}_1, \cdots, \mathbb{Y}_{N_q}\}$. Due to the temporal and spatial variations within the timeseries signal, we propose a likelihood function that takes into account the convolution of the spatial location and intensity of the Digital Twin methane enhancements relative to the measurement-based enhancements. The formulation is written as

$$\phi_{j,k} = \sum_i^N \exp\left\{-\frac{(y_{m,k} - y_{j,i})^2}{2\sigma_y^2}\right\} \exp\left\{-\frac{d_{k,i}^2}{2\sigma_d^2}\right\}, \tag{9.23}$$

where $\phi_{j,k}$ represents the general weight for the j-th DT at the k-th measurement time step, $y_{m,k}$ is the measured enhancement at the k-th time step, and $d_{k,i}$ is the Euclidean distance between the spatial location of $y_{m,k}$ and $y_{j,i}$. The σ_y and σ_d values can be tuned to accept signals that are more or less close to the true enhancement signal and spatial location, respectively. We then define the sum of the weights for the j-th DT to be ϕ_j, such that,

$$\phi_j = \sum_k^N \phi_{j,k}. \tag{9.24}$$

The weight vector for the likelihood can then be estimated as,

$$w_j = \frac{\phi_j}{\|\Phi\|_1}, \quad \Phi = [\phi_1, \cdots, \phi_{N_q}]. \tag{9.25}$$

The values of $\mathbf{w} = [w_1, \cdots, w_{N_q}]$ given its source location $\mathbb{X}_q$ are a good approximation of $p(\mathbb{D}|\Theta)$, since they use both the data, $\mathbb{D}$, and the current estimate of the parameters, Θ. If any of the w_j's are close to zero, we can interpret this as a low likelihood or probability that the source location is at the j-th point.

Then using the prior and likelihood we can formulate the posterior update for one iteration as,

$$p(\Theta|\mathbb{D}) \propto p(\mathbb{D}|\Theta)p(\Theta). \tag{9.26}$$

Since the prior is uniform, the posterior is proportional to the likelihood. Additionally, we can repeat the iteration step till we compute the l-th posterior weight estimate, $\mathbf{w}^{(l)}$. Then we can take the expectation of the weight vectors to update the weight estimate,

$$\hat{\mathbf{w}} = \frac{\frac{1}{N_l}\sum_{l=1}^{N_l} \mathbf{w}^{(l)}}{\|\frac{1}{N_l}\sum_{l=1}^{N_l} \mathbf{w}^{(l)}\|_1}. \tag{9.27}$$

Once the weight estimate are determined the source rates can be updated, such that, for each $q_j = w_j Q$, and $\sum_{j=1}^{N_q} q_j = S$ (from Assumption 1).

Refinement Step: After the first posterior computation, we propose a refinement step to improve the resolution and overall accuracy of the solution. Let us define a threshold value, $\tau_q > 0$, such that the values that do not meet this threshold (i.e. $q_j \leq \tau_q$) are removed from the set of possible source locations and source rates in Θ. The remaining source locations and rates are then refined for the next iterative step of $p(\Theta)$, such that the j-th grid location is broken into quadrants and four new locations replace the previous j-th source location (see illustration in Fig. 9.12 for the first iteration). This process was motivated by the multi-target search problem using the probabilistic quadtree (see [3]). Alternative approaches to refine the spatial grid space can utilize spatial kriging of the weights before removing the source rates and locations that do not meet the threshold. Ultimately, the iterative refinement methodology can be repeated until the desired resolution is reached – see the brief summary in Algorithm 12.

9.3.2 Experiment

To test the DT framework and iterative refinement method for the landfill emission source determination problem, we first need to generate a terrain map of a desired landfill. From previous flights (shown as a case study in [10]), we carried out multi-rotor-based methane quantification flights at the Merced County Regional Waste Management (MCRWM) site located in Merced, CA, USA. The task was to quantify emissions coming from the adjacent side of an active cell, where a new cell was currently being excavated. For simulation purposes, we focused on this area as our simulated unknown area source emission. To create the terrain map, a raster scanning path was used in Google Earth to extract terrain height location data. The resulting

---

Algorithm 1: The iterative refinement method

Data: Given $\mathbb{D}$ and $P(\Theta)$, collect $\mathbb{Y}_j$'s

Result: Source rate/location estimates via $P(\Theta|\mathbb{D})$

1 **while** *Desired resolution is not met* **do**
2   compute DT forward models;
3   collect measurement data, $\mathbb{Y}_j$;
4   compute weight estimates, $\mathbf{w}^{(l)}$;
5   **if** *iteration==max iterations* **then**
6   Compute $\hat{\mathbf{w}}$;
7   Update Θ using threshold;
8   Refine grid and source rate locations;
9   Re-initialize $P(\Theta)$;
10   iteration = 0;
11   **else**
12   iteration++;

KML file was used in conjunction with kriging to create the uniformly spaced terrain map, $\mathbb{T}$.

To sample the airspace above the MCRWM site, we proposed a takeoff location just to the east of the active cell at MCWRM along an access road. The flight pattern we chose here is designed along the perimeter of the active site, referred to as the perimeter flight pattern (PFP). This pattern is an adjustment to the typical cylindrical flux plane (CFP) flight trajectory in that it flies along geographic boundaries and obstructions (creating a closer flight path to the potential emission source). The CFP in general encapsulates the source through a circular flight trajectory capturing the flux entering and leaving through the CFP. See Fig. 9.12 for an illustration of the flight path.

Given that flight path encloses the active cell, the a priori source distribution locations are initialized across the entire active cell in an uniform manner (≈ 100 m spacing). Additionally, the physical system and DT forward simulations (given the true and estimated source distribution locations and weights) were run for several minutes before starting the flight trajectory path and taking measurement data. This was done to ensure the plume was fully developed. Since the simulation and real-world experiments typically have stochastic components, we chose to simulate four complete flights of the physical system. For each iterative refinement level, the DT system is simulated 10 times. A comparison between all combinations of physical and DT system are carried out to estimate the new weights, based on (9.27). The simulation experiment consisted of four refinement levels. The grid spacing of refinement level 1 started at 100 m and was reduced to 50 m, 35 m, 20 m, and 10 m, for levels 2–4, respectively.

The true source distribution was created using a grid spacing of 25 m and a Gaussian function with multiple peaks known as Gaussian sums,

$$f(\mathbf{x}; \theta) = \frac{1}{2\pi \sigma_{x_1} \sigma_{x_2}} \sum \exp\{\frac{-(\mathbf{x}_1 - \mu_1)^2}{2\sigma_{x_1}^2} - \frac{(\mathbf{x}_2 - \mu_2)^2}{2\sigma_{x_2}^2}\}, \qquad (9.28)$$

where the parameter $\theta = [\mu_1, \mu_2, \sigma_{x_1}, \sigma_{x_2}]$. To test the algorithm we tested a two peak area source distribution with parameters, $\theta_1 = [445, 470, 25, 25]$ and $\theta_2 = [445, 550, 25, 25]$ (see Fig. 9.12).

9.3.3 Results

After the first set iteration steps, the weight estimate highlights an area near the true area source rate. Following the iteration steps, the weight estimate still encompasses the location near the true area source but also includes locations in the direction of the wind (see Fig. 9.13). Additionally, for each DT forward model run, the measurement signal characteristics can be compared with the true measurement signal (see Fig. 9.14). After the fourth iterative refinement, the area source resolution of the DT has reached 10 m. In Fig. 9.15, we can see the final area source distribution compared to the original source. The error of the center of both sources are shown to be around 106 m. It can also be observed that the solution is biased toward the pre-dominant wind direction.

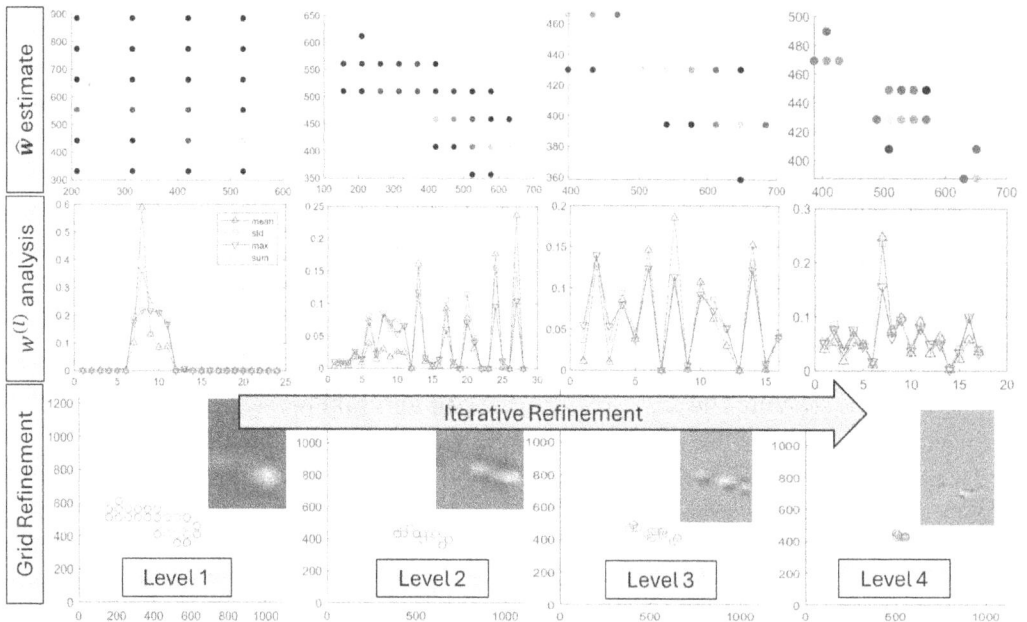

FIGURE 9.13 The weight estimate, analysis, and grid refinement results are shown from left to right, where the current iteration guess is shown on the top, and based on the analysis (middle) the resulting refined grid points are shown on the bottom. (reused from [14] with permission from ©IEEE)

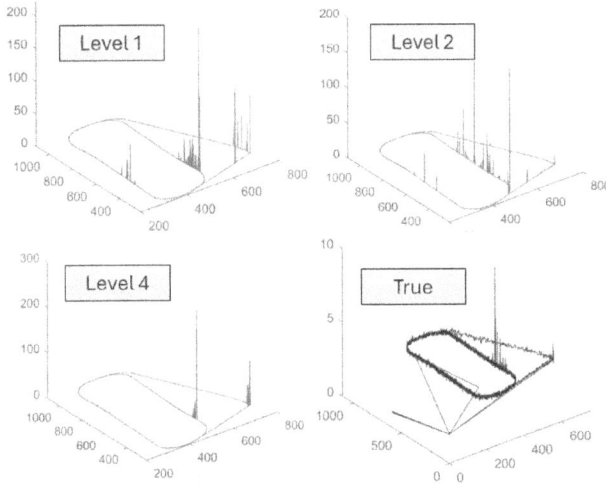

FIGURE 9.14 The characteristics of the time-series signals are shown at different refinement levels against the true measured signal (the different shades represent the different point source DT signals). It can be observed that the solution behavior gets qualitatively closer to the true signal through each iterative refinement.(reused from [14] with permission from ©IEEE)

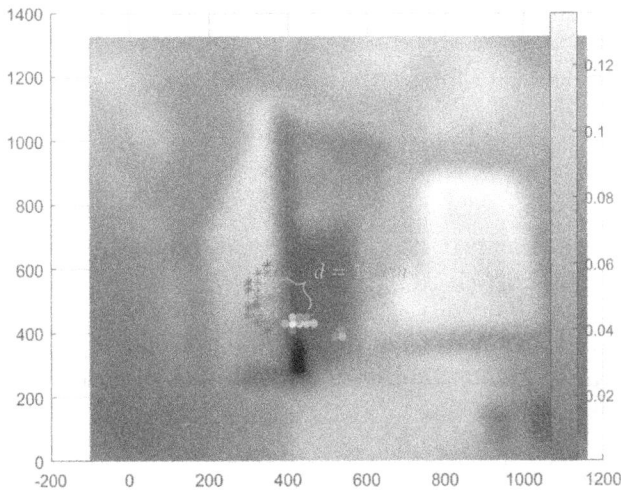

FIGURE 9.15 The true area source distribution is shown with the asterisk symbol and the iterative refinement solution is represented with the solid dots. The fraction of the source rate attributed to a specific spatial point is indicated by the shade intensity. It can be observed that the iterative refinement distribution is primarily represented by a few points in close proximity to the true distribution. (reused from [14] with permission from ©IEEE)

Pause and Reflect

As we have just seen, DTs can be used to make our measurement systems smarter, improving where and how to best sense. However, there is still more work to be done (e.g. behavior matching or forward model methodologies). Based on your knowledge of the subject, what are the ways that we can improve on this framework?

9.4 CHAPTER SUMMARY

Digital Twins are more than just a tool to model the environment, they can help improve the way we sense. In this chapter we showcased three examples of how digital twins can be leveraged to extract more from environment than just measurement data. The first example, gives insight into how multiple sensors can be outfitted across a landscape to better detect the emission source, but more importantly better estimate the location. The results prompt research questions, such as, how many sensors are needed, and where to place the sensors to maximize detection while minimizing cost? The second example, highlights the value of the empirical Gramian and how DTs can be used to improve how we autonomously steer a swarm of drones for quantifying an emission source, but also improving the performance. Deviating from traditional path planning strategies to include dynamically moving formation control, the field measurements are providing a level of persistence of excitation in the measurement signal to help estimate the emission source. The last example showcases the problem of estimating an area source from data gathered downwind of the source. This method is critical for efficiently and accurately estimating the emissions of area sources while mapping them for inspection and repair.

Bibliography

[1] Rihab Abdul Razak, Srikant Sukumar, and Hoam Chung. Scalar field estimation with mobile sensor networks. *International Journal of Robust and Nonlinear Control*, 31(9):4287–4305, 2021.

[2] Jianxiong Cao, YangQuan Chen, and Changpin Li. Multi-UAV-based optimal crop-dusting of anomalously diffusing infestation of crops. In *Proc. of the 2015 American Control Conference (ACC)*, pages 1278–1283. IEEE, 2015.

[3] Stefano Carpin, Derek Burch, and Timothy H Chung. Searching for multiple targets using probabilistic quadtrees. In *2011 IEEE/RSJ International Conference on Intelligent Robots and Systems*, pages 4536–4543. IEEE, 2011.

[4] YangQuan Chen, Zhongmin Wang, and Jinsong Liang. Actuation scheduling in mobile actuator networks for spatial-temporal feedback control of a diffusion process with dynamic obstacle avoidance. In *IEEE International Conference Mechatronics and Automation, 2005*, volume 2, pages 752–757 Vol. 2, 2005.

[5] YangQuan Chen, Zhongmin Wang, and Jinsong Liang. Optimal dynamic actuator location in distributed feedback control of a diffusion process. *International Journal of Sensor Networks*, 2(3-4):169–178, 2007.

[6] YangQuan Chen, Zhongmin Wang, and K.L. Moore. Optimal spraying control of a diffusion process using mobile actuator networks with fractional potential field based dynamic obstacle avoidance. In *Proc. of the 2006 IEEE International Conference on Networking, Sensing and Control*, pages 107–112, 2006.

[7] Michael A Demetriou. Inducing persistence of excitation through sensor motion in the adaptive estimation of spatial fields. In *2022 American Control Conference (ACC)*, pages 1673–1678. IEEE, 2022.

[8] Jay A Farrell, John Murlis, Xuezhu Long, Wei Li, and Ring T Cardé. Filament-based atmospheric dispersion model to achieve short time-scale structure of odor plumes. *Environmental Fluid Mechanics*, 2(1-2):143–169, 2002.

[9] Christian Himpe. emgr—the empirical Gramian framework. *Algorithms*, 11(7):91, 2018.

[10] Derek Hollenbeck. *Digital Twin Enabled Collective Sensing and Steering for Source Determination Problems*. PhD thesis, University of California, Merced, 2023.

[11] Derek Hollenbeck and YangQuan Chen. Characterization of ground-to-air emissions with sUAS using a digital twin framework. In *Proc. of the 2020 International Conference on Unmanned Aircraft Systems (ICUAS)*, pages 1162–1166. IEEE, 2020.

[12] Derek Hollenbeck and YangQuan Chen. A more optimal stochastic extremum seeking control using fractional dithering for a class of smooth convex functions. *IFAC-PapersOnLine*, 53(2):3737–3742, 2020.

[13] Derek Hollenbeck and YangQuan Chen. Digital twin behavior matching of gas plumes using a fixed sensor mesh and dynamic mode decomposition. In *Proc. of the 2021 International Design Engineering Technical Conferences and Computers and Information in Engineering Conference*, volume 85437, page V007T07A005. American Society of Mechanical Engineers, 2021.

[14] Derek Hollenbeck and YangQuan Chen. Towards a digital twin framework for landfill emission source determination using hybrid VTOL fixed-wing: A simulation study. In *2024 20th IEEE/ASME International Conference on Mechatronic and Embedded Systems and Applications (MESA)*, pages 1–7. IEEE, 2024.

[15] Derek Hollenbeck, Demitrius Zulevic, and YangQuan Chen. A modified near-field Gaussian plume inversion method using multi-sUAS for emission quantification. In *2022 International Conference on Unmanned Aircraft Systems (ICUAS)*, pages 1620–1625. IEEE, 2022.

[16] Dariusz Ucinski and Maciej Patan. Sensor network design for the estimation of spatially distributed processes. *International Journal of Applied Mathematics and Computer Science*, 20(3):459, 2010.

[17] Jairo Viola, Derek Hollenbeck, Carlos Rodriguez, and YangQuan Chen. Fractional-order stochastic extremum seeking control with dithering noise for plasma impedance matching. In *Proceedings of the 2021 Conference on Control Technology and Applications*. IEEE, 2021.

CHAPTER 10

Conclusions and Best Practices

Throughout this book, we investigated the emission source determination problem, first, in the context of detection, localization, and quantification with sUAS, and second, in the context of Digital Twins, the sensor placement and steering problem, and smart sensing. Here we provide a general summary of the topics covered and learned in this book, a list of best practices from lessons learned in case studies, some future research directions to explore to improve and expand on this work, and lastly, a breakdown of how the MOABS/DT code is built, including some example implementation codes.

10.1 CONCLUSIONS

In Chapter 1, we introduced the methane sensing problem and the importance of making measurements to efficiently manage the total amount of emissions, and thereby reduce the effects of methane on the environment. Then we overviewed the different possible methane emission sources that can be likely seen in practice. Lastly, we looked into the different measurement modes and technologies associated with them. This was important as it has implications in following chapters.

In Chapter 2, we investigated emission source detection from the sUAs point of view. This started with overviewing the different platforms (multirotor, fixed-wing, and hybrid VTOL), and how sensor can be potentially integrated onto those platforms. The sensor integration and placement on the aircraft are critical for detection and depend on the type of emission source targeted to detect. We ended on discussing the importance of understanding how these measurement systems can have internal detection limits and when applied to intermittent sources or under different weather conditions and flight conditions, the probability of detection changes. If the operator wants to sufficiently say they have surveyed a region and did not detect an emission source, they need to look at probability of detection.

In Chapter 3, we discussed the emission source localization problem. Beginning with a general definition and introducing the different localization types (direct search, grid search, inference, control volume, and back trajectory), we still found there are difficulties in applying these methods. Challenges include, atmospheric variability, complex terrain, pin-pointing, or operating constraints (e.g. standoff distances).

In Chapter 4, we explored a large body of work surrounding emission source quantification. This began with the different types of quantification: simulation, optimization, mass-balance, imaging, and correlation-based approaches. We learned that there are several approaches within mass-balance and some in optimization that can be directly applied with sUAS (or already have). Each of these methods need to be applied slightly differently depending on which sensing mode and technology used. Additionally, the sUAS platform type comes into effect when considering minimum detection level. For example, fixed-wing sUAS travel much faster than multirotor platforms, resulting in larger spacings between sample points (given the same sample rate). This is key when thinking about detection probabilities that we learned in Chapter 2. Lastly, we provided an overview of some techniques for processing data and controlling multi-sUAS systems (which is introduced later) before ending with an assessment summary of the quantification methods discussed (in terms of cost, complexity, and precision).

In Chapter 5, we showcased several case studies within detection, localization, and quantification themes from previous chapters. The experiments led to many observations and lessons learned. Most of which are related to atmospheric conditions and the variable wind directions. These experiments ranged from rural, to urban, to simulated neighborhoods, to simulated oil and gas scenarios. We were also able to explore quantification with area sources, such as those found in landfills, and in the arctic permafrost.

In Chapter 6, we began the second half of the book, which targets embedding smartness into the source emission determination problem with sUAS. This started with the introduction of the Digital Twin (DT), and how we can numerically represent them. We explored the concept of behavior matching and the difficulties therein. For example,

$$\hat{\theta} = \min_{\theta} \|\mathcal{F}(\theta) - y\|, \tag{10.1}$$

where $\mathcal{F}$ might be a numerical solver, closed-form model, or neural operator. We ended the chapter by deriving the MOABS/DT platform, which is the work horse for the following chapters.

In Chapter 7, we showcased four different case studies using the MOABS/DT platform. The first focused on behavior matching the data by directly comparing the output of the model to that of the observed data using a pattern search optimization. This was informative of just how hard behavior matching can be for complex system, much less in real-time (this is the ultimate goal). We then explored how we can test methodologies using the newly behavior matched DT, such as with the Mittag-Leffler weighted time dependent kernel DM+V method. While the case study was not conclusive, it did still shed light on the importance of atmospheric weather conditions. The next two case studies developed, for the first time, a multi-sUAS method for quantifying the emission source rate. The results were then compared with single and multi-sUAS methods in the last case study, which showed that the mod-NGI-CVT method was better at many downwind distances. For large multi-sUAS teams, the method was far superior to the others.

In Chapter 8, we expand on the use of DTs by introducing the concept of smart sensing, along with, the sensor placement and steering problem, and observability Gramian. The primary challenge is related to the ability to analyze a model efficiently with respect to the parameters in an optimal way. This is typically looked at as an inverse problem or a parameter estimation problem. In either case, regression techniques can be applied but still require efficient forward model computation. For example,

$$\hat{\theta} = \min_{\theta} \|\mathcal{F}(\theta) - y\| + \lambda \mathcal{R}(\theta), \tag{10.2}$$

where the term $\mathcal{R}$ can be a regression term, such as, $\|\theta\|_{\ell_2}$ and $\|\theta\|_{\ell_1}$, or a residual (also referred to as a misfit term) to constrain the fit to the governing PDE equations. The is especially the case for the sensor placement and steering problem, where the mobile sensor system needs to find the optimal path to improve parameter estimates. It is worth noting that within these frameworks, the sensor needs to be persistently exciting the states of the system (i.e. the plume) in order to meet the sufficient richness condition and estimate the parameters. Therefore, having a 'well matched' DT is necessary to ensure we meet this condition.

In Chapter 9, we showcased three case studies of smart sensing. The first example, includes utilizing the DT to estimate the location of an emission source through online behavior matching based on a set of distributed sensors. This illustrates a use case for continuous emission monitoring but also can be extended to the mobile sensor case. The second example, introduces a continuous CVT model for inducing persistence of excitation into the formation control algorithm for quantifying emissions using the mod-NGI method. The plane controller was informed of the most observable direction by the empirical Gramian, which illustrates another useful feature of DTs. The last example focused on using the DT to solve an area source inverse problem. The sUAS was tasked with quantifying a large area emission source by using a cylindrical flux plane. The observed data was used with the DT, to develop a model based on an iterative refinement framework. The result was an estimate of the area source distribution near the true source distribution. Although not exact, the method did show some bias in the direction of the wind.

10.2   BEST PRACTICES

Sensor Selection and Placement – The best practice of sensor selection and placement can be broken into a few steps:

1. Determine with what kind of emission sources are to be detected and the minimum sensitivity threshold.

   - Large emission rates correlate to larger concentrations and less sensitivity requirements.

   - Small emission rates correlate to small concentrations and higher sensitivity requirements.

2. Characterize the sensor in terms of the response time, linearity, drift, etc.

   • Utilize fume hood tests or wind-tunnel tests to isolate conditions and verify manufacturers specifications.

   • There is a typical trade-off between good sensor performance values and cost.

   • Another trade-off exists between sensor performance and sensor weight. Lightweight sensors typically have a performance decrease compared to state-of-the-art (SOTA) benchtop and portable units.

3. Determine a mounting location that is compatible with your sensor type and sensing modality/technology.

   • For example, an in situ sensor may be best placed in a boom-mounted (or joust) configuration.

   • a path-integrated sensor may be best suited for a bottom-mounted configuration.

4. Verify the sensor integration has bias or vehicle induced disturbances.

   • Utilize smoke tests to visually determine bias and disturbances, e.g. [22] or [9].

   • Utilize controlled release tests or field experiments to measure bias and disturbances directly.

   • Compare results with those found in characterization steps.

Source Localization Techniques – the best practice for source localization depends on the sensor selection and integration step, along with the expected emission types and environmental conditions. Here is a list of things to consider:

1. Verify the weather conditions support the sUAS capabilities and are sufficient for the method you intend to apply;

   • Check if wind speed and direction are consistent and within the stable atmospheric class conditions (or close to it).

   • Wind speed (or gusts) do not exceed sUAS manufacturers recommended airworthiness values.

2. Make a list of possible emission source locations and types.

   • Equipment sources: e.g. compressor stations, well heads, pump jack, tanks, thief hatch, underground pipes, etc.

   • Natural sources: e.g. sewage, landfills, wetlands, leachate ponds, etc.

3. Conduct an initial survey of the site.

- Utilize the control volume method, such that when a source is detected downwind, but not upwind, we can narrow the localization to a specific region of the site.

4. Revisit likely regions for leak verification.

   - Utilizing some thorough back trajectories or upwind survey regions [23], a source location can be determined.

   - The extent of the localization precision is typically on the order of meters, but if higher precision is needed, an optimization-based approach may be sought.

   - In some cases, the observed leak location may not be the actual source location. For example, an underground pipe may manifest itself in a different location than where it is broken. In some complex terrains, a grid-based search may be more suitable, as wind patterns are more unpredictable.

Quantification Strategy – the best practice for quantification may depend on several factors. In general, it is assumed the leak or emission source has been sufficiently localized, such that, a downwind flight path can be designed. Things to consider for quantification are:

1. Current weather conditions:

   - Wind speeds above 2m/s and below the airworthiness limits of the sUAS.

   - Wind direction is predominately coming from one direction (i.e. stable).

   - Temperatures or humidity conditions do not exceed the operating conditions of the sensor.

2. Sensor and downwind sampling distance:

   - Understand the sensor's sample rate and sensitivity to select an appropriate downwind distance.

   - The behavior of the methane signal will change as a function of downwind distance. The farther downwind, the more the signal becomes 'well mixed' (>100m–1000m)

   - Measuring close to the source will result in higher concentration signal but the signal will be highly localized and turbulent in nature (e.g. spikey).

   - Find a trade-off between the sensor's sensitivity limitations and the well-mixed condition.

3. Flight path and method:

   - Given a mass balance method [2, 1, 5, 4, 7, 8, 15, 16, 17, 19, 20, 27, 26, 25, 3] or Near-field Gaussian plume model inversion (NGI) [18, 10], a vertical flux plane can be flown.

- Given a Gauss Divergence Theorem approach [6], a cylindrical flux plane can be flown.
- Based on your sUAS flight time, find a suitable height change between transects, such that the flux plane is thoroughly scanned.
- Make sure the final altitude reaches a clean air fetch and methane levels are at background (this may vary from location to location).

Digital Twin Integration – The best practices of digital twin integration depend on the level of fidelity desired. In some scenarios a low fidelity model can be used to represent the DT (e.g the Gaussian plume model), however, there are many cases where this will not suffice. Here are some things to consider:

1. Real- or near real-time dispersion modeling. Integrating the DT into the measurement system's hardware requires a fast running model and is a technical bottle neck for online level 3 DT.

2. If using existing code, make sure you can access source code for use in behavior matching optimization as well as deployment on the edge or GCS devices. A list of EPA approved methods can be found online[1] and other numerical approaches in [11].

3. Most of the practical DT integrations will be a level 2 DT at best. Thus, there is still more work to be done on our ability to compute high fidelity dispersion models faster than real-time.

10.3 FUTURE EFFORTS

This book only delivers an introduction at best into the smart sensing framework and environmental sensing of methane with DTs. As a result, there is plenty of research left to do. Here is a list of potential research directions that would complement this work:

1. Investigation into the integration of low-cost (or relatively low-cost) sensors with high-resolution simulators, to reduce the entry barrier for emission measurements.

2. Investigation and development into the feasibility of hybrid approaches that combine machine learning [12, 24] and physical models to improve the computational efficiency and fidelity.

3. Understanding the efficacy of using partial integral algebra to represent the smart sensing framework and dispersion modeling from a control theoretic point of view [21].

4. Investigating and developing behavior matching optimization algorithms that work with hybrid modeling approaches, e.g. behavioral systems theory [14, 13], forward-backward dispersion modeling, or federated learning.

---

[1]https://www.epa.gov/scram/air-quality-dispersion-modeling

5. Developing fast-running models to improve scalability for real-time estimation on embedded systems (i.e. edge computing).

6. Investigation into the transferability of learned models across sites with different configurations, atmospheric conditions, and emission types.

7. Evaluating different multi-sUAS path planning strategies for coordinated smart sensing with collision avoidance capabilities.

8. Developing a full integration with an operational digital twin (in the Level 3 sense) for real-time smart sensing.

10.4   CODE

The code for the MOABS/DT platform, derived in Chapter 6, is broken down into the class architecture and example code to get started. The code can be found in the Github[2].

10.4.1   Architecture

As DTs need to be behavior matched and/or conditions may change, it was important to design the MOABS/DT platform to be object-oriented. In traditional modeling approaches, the simulation or computation is done in a procedural way. The disadvantage is that the code needs to be modified for different scenarios. When MOABS/DT was written, we wanted to design the code, in such a way, that the simulation can be modified without major changes to the source code. This flexibility is useful for simulating multiple emission sources or simulating mobile emission sources, or even adding or removing emission sources during the simulation. Since all of the code is embedded within the object-oriented class it is quite versatile. In the next few paragraphs, we will overview the architecture of the classes.

In Fig. 10.1, we can see that the DT begins with an Environment Class. The Environment Class acts as a container for the other classes and in essence acts as the DT environment (hence the name). The other classes act as plugins to the Environment Class and are not restricted to be a single instance of the class (e.g. they can be vectors).

10.4.2   Hello World Example

To simulate a plume within the MOABS/DT platform architecture, the user needs to instantiate the different different classes using their constructors. If default values are not desired, the user needs to set the particular class variables using a struct variable. The simplest case is a simulation with no Surface Class, no Filament Class, and no Planner Class (i.e. just simulating the wind). This is typically done using a 'settings' script.

---

[2]https://github.com/dhollenbeck1/moabs

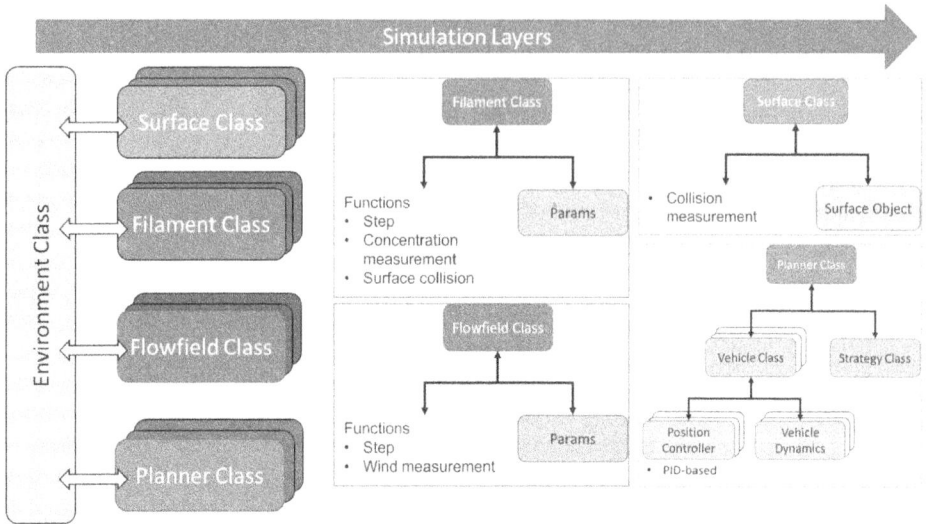

FIGURE 10.1   The MOABS/DT Class Architecture.

```
%% Example 0 Settings
% -- set simulation params
simparams.N = 60;
simparams.modnum = 20;

% -- create param structs for initialization
ff_params              = struct;
ff_params.dim          = 3;
ff_params.uv0          = [1,0];
ff_params.alpha         = 0.3;
ff_params.abG          = [0.005,0.02,1];
```

The main code is then first setup with a desired working directory (wk_dir) before initilizing the the simulation objects and the main loop iterations (based on the Environment step frequency – default is 20 Hz). The simple flowfield example is given in the following code boxes. The first is the header of the main loop.

```
%% Example 0
% This script runs a MOABS/DT simulation of the
    flowfield only.

% -- INITIALIZE SETTINGS AND ADD FOLDER PATHS
clc, clear
wkdir_path = <your path>;
addpath([wkdir_path,'Classes\'])
addpath([wkdir_path,'Utility\'])
example0_settings
```

```
PLOTRESULTS = 1;

% -- INITIALIZE SIMULATION OBJECTS
ff = flowfieldClass(ff_params);

% -- initialize environment
env = environmentClass();
env = env.addFlowfield(ff); clear ff;
% env.az = 90; env.el = 0;

% -- main loop settings
n = simparams.N*env.Fs;
```

If you notice that the Environment can take the flowfield object and then we can delete it from memory. The view of the output can also be adjusted. This is important when we have 3D representations but can also be adjusted temporally for added effects. The next code block gives an example of the main loop itself. This block is very simplified but can be expanded to include timing information or intermediate steps.

```
% -- MAIN LOOP
figure(1)
for i=1:n
    % -- step environment
    env = env.step();

    % -- plot results
    if mod(i,2*env.Fs)==0 && PLOTRESULTS
        clf
        env.plotff(1)
        shg
    end
end
```

By varying the a, b, and G parameters found in ff_params.abG in the settings script, the level of meandering can be increased or decreased.

10.4.3 2D plume example

To implement the example of a plume dispersing in an advection field, we need to add a few lines to the settings file (see code box below).

```
plume_params             = struct;
plume_params.p0          = [40 100 0];
plume_params.Collisions = 'False';
```

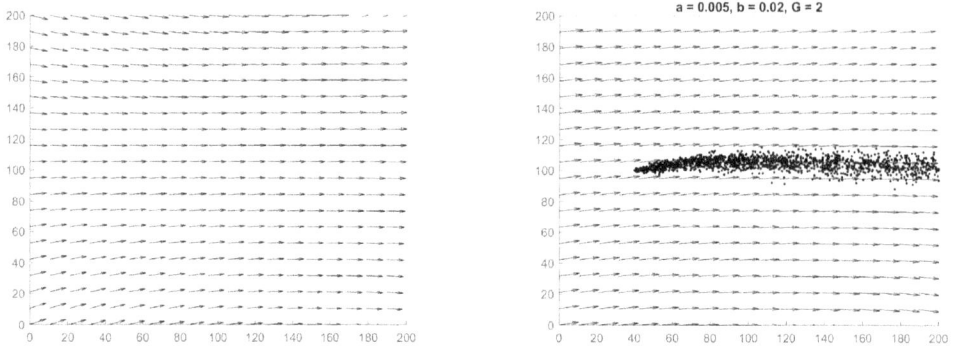

FIGURE 10.2 (left) MOABS/DT wind field example output and (right) a simple 2D plume example output.

Then, by adding several lines to the main script (where it says initialize environment) we can add the plume class to the environment object and then simulate the plume dispersing in the environment, subject to the wind field parameters.

```
plume = filamentClass(plume_params);
env = env.addPlume(plume); clear plume;
```

To visualize the plots we need to add the plotting commands to the environment object (see Fig. 10.2).

```
% -- plot results
    if mod(i,modrate)==0 && PLOTRESULTS
        clf
        if isempty(env.terrain)==0
            env.plotterrain(0)
        end
        env.plotff(1)
        env.plotplume(1)
        shg
    end
```

The terrain can also be plotted as well if it is desired. This requires the Terrain Class object to be attached to the Environment Class object (e.g. 'env' in this case). For more implementation details, see the Github Wiki[3], which will serve as a living document as the DT model develops and new applications are derived.

---

[3]https://github.com/dhollenbeck1/moabs/wiki

10.5 DATA

The inclusion of data sets in research is a solid way to improve the validity of contributions and increase the reproducibility. However, in this book, not all the data sets are available for sharing. Here we list a few data sets used in Chapter 5 that are published on the Data Dryad[4] (see list below).

10.5.1 Data set title list

- 2017 Methane Emission Quantification and Localization Controlled Release Test at the Merced Vernal Pools and Grassland Reserve with sUAS

- 2017 Methane Emission Detection and Localization Controlled Release Test at the PG&E Livermore Facility with sUAS

- 2018 Methane Emission Quantification Controlled Release Test at the Merced Vernal Pools and Grassland Reserve with sUAS

Bibliography

[1] David Allen, Shannon Stokes, Erin Tullos, Brendan Smith, Scott Herndon, and Bradley Flowers. Field trial of methane emission quantification technologies. In *Proc. of the SPE Annual Technical Conference and Exhibition.* OnePetro, 2020.

[2] Grant Allen, Peter Hollingsworth, Khristopher Kabbabe, Joseph R Pitt, Mohammed I Mead, Samuel Illingworth, Gareth Roberts, Mark Bourn, Dudley E Shallcross, and Carl J Percival. The development and trial of an unmanned aerial system for the measurement of methane flux from landfill and greenhouse gas emission hotspots. *Waste Management*, 87:883–892, 2019.

[3] M Bourn, G Allen, P Hollingsworth, K Kababbe, P I Williams, H Ricketts, J R Pitt, and A Shah. The development of an unmanned aerial system for the measurement of methane emissions from landfill. *Sixteenth International Waste Management and Landfill Symposium*, (October 2017), 2018.

[4] Maria Obiminda L Cambaliza, Jean E Bogner, Roger B Green, Paul B Shepson, Tierney A Harvey, Kurt A Spokas, Brian H Stirm, Margaret Corcoran, Detlev Helmig, and Armin Wisthaler. Field measurements and modeling to resolve m^2 to km^2 CH_4 emissions for a complex urban source: An Indiana landfill study. *Elementa: Science of the Anthropocene*, 5, 2017.

[5] MOL Cambaliza, PB Shepson, DR Caulton, B Stirm, D Samarov, KR Gurney, J Turnbull, KJ Davis, A Possolo, A Karion, et al. Assessment of uncertainties of an aircraft-based mass balance approach for quantifying urban greenhouse gas emissions. *Atmospheric Chemistry and Physics*, 14(17):9029–9050, 2014.

---

[4]https://datadryad.org – search using the data set titles.

[6] Stephen Conley, Ian Faloona, Shobhit Mehrotra, Maxime Suard, Donald H Lenschow, Colm Sweeney, Scott Herndon, Stefan Schwietzke, Gabrielle Pétron, Justin Pifer, et al. Application of Gauss's theorem to quantify localized surface emissions from airborne measurements of wind and trace gases. *Atmospheric Measurement Techniques*, 10(9):3345–3358, 2017.

[7] James L France, Prudence Bateson, Pamela Dominutti, Grant Allen, Stephen Andrews, Stephane Bauguitte, Max Coleman, Tom Lachlan-Cope, Rebecca E Fisher, Langwen Huang, et al. Facility level measurement of offshore oil and gas installations from a medium-sized airborne platform: method development for quantification and source identification of methane emissions. *Atmospheric Measurement Techniques*, 14(1):71–88, 2021.

[8] Derek Hollenbeck and YangQuan Chen. Characterization of ground-to-air emissions with sUAS using a digital twin framework. In *Proc. of the 2020 International Conference on Unmanned Aircraft Systems (ICUAS)*, pages 1162–1166. IEEE, 2020.

[9] Derek Hollenbeck, Madoka Oyama, Andrew Garcia, and YangQuan Chen. Pitch and roll effects of on-board wind measurements using sUAS. In *Proc. of the 2019 International Conference on Unmanned Aircraft Systems (ICUAS)*, pages 1249–1254. IEEE, 2019.

[10] Derek Hollenbeck, Demitrius Zulevic, and YangQuan Chen. A modified near-field Gaussian plume inversion method using multi-sUAS for emission quantification. In *2022 International Conference on Unmanned Aircraft Systems (ICUAS)*, pages 1620–1625. IEEE, 2022.

[11] Nicholas S Holmes and Lidia Morawska. A review of dispersion modelling and its application to the dispersion of particles: an overview of different dispersion models available. *Atmospheric Environment*, 40(30):5902–5928, 2006.

[12] George Em Karniadakis, Ioannis G Kevrekidis, Lu Lu, Paris Perdikaris, Sifan Wang, and Liu Yang. Physics-informed machine learning. *Nature Reviews Physics*, 3(6):422–440, 2021.

[13] Ivan Markovsky and Florian Dörfler. Behavioral systems theory in data-driven analysis, signal processing, and control. *Annual Reviews in Control*, 52:42–64, 2021.

[14] Thabiso M Maupong and Paolo Rapisarda. Data-driven control: A behavioral approach. *Systems & Control Letters*, 101:37–43, 2017.

[15] Randulph P Morales, Jonas Ravelid, Killian P Brennan, Béla Tuzson, Lukas Emmenegger, and Dominik Brunner. Estimating local methane sources from drone-based laser spectrometer measurements by mass-balance method. In *EGU General Assembly Conference Abstracts*, page 14778, 2020.

[16] Arvind P Ravikumar, Sindhu Sreedhara, Jingfan Wang, Jacob Englander, Daniel Roda-Stuart, Clay Bell, Daniel Zimmerle, David Lyon, Isabel Mogstad, Ben Ratner, et al. Single-blind inter-comparison of methane detection technologies–results from the Stanford/EDF Mobile Monitoring Challenge. *Elementa: Science of the Anthropocene*, 7, 2019.

[17] Maximilian Reuter, Heinrich Bovensmann, Michael Buchwitz, Jakob Borchardt, Sven Krautwurst, Konstantin Gerilowski, Matthias Lindauer, Dagmar Kubistin, and John P. Burrows. Development of a small unmanned aircraft system to derive CO_2 emissions of anthropogenic point sources. *Atmospheric Measurement Techniques*, 14(1):153–172, 2021.

[18] Adil Shah, Grant Allen, Joseph R Pitt, Hugo Ricketts, Paul I Williams, Jonathan Helmore, Andrew Finlayson, Rod Robinson, Khristopher Kabbabe, Peter Hollingsworth, et al. A near-field Gaussian plume inversion flux quantification method, applied to unmanned aerial vehicle sampling. *Atmosphere*, 10(7):396, 2019.

[19] Adil Shah, Grant Allen, Hugo Ricketts, Joseph Pitt, and Paul Williams. Methane flux quantification from lactating cattle using unmanned aerial vehicles. *EGU General Assembly*, 20:7655, 2018.

[20] Adil A Shah. Methane Flux Quantification Using Unmanned Aerial Vehicles. *Diss., University of Manchester*, 2020.

[21] Sachin Shivakumar, Amritam Das, Siep Weiland, and Matthew Peet. Extension of the partial integral equation representation to GPDE input-output systems. *IEEE Transactions on Automatic Control*, 2024.

[22] Brendan Smith, Garrett John, Brandon Stark, Lance E Christensen, and YangQuan Chen. Applicability of unmanned aerial systems for leak detection. In *Proc. of the 2016 International Conference on Unmanned Aircraft Systems (ICUAS)*, pages 1220–1227. IEEE, 2016.

[23] Witenberg SR Souza, Alexander J Hart, Benedito JB Fonseca, Mansour Tahernezhadi, and Lance E Christensen. A framework to survey a region for gas leaks using an unmanned aerial vehicle. *IEEE Access*, 12:1386–1407, 2023.

[24] Shashank Subramanian, Peter Harrington, Kurt Keutzer, Wahid Bhimji, Dmitriy Morozov, Michael W Mahoney, and Amir Gholami. Towards foundation models for scientific machine learning: Characterizing scaling and transfer behavior. *Advances in Neural Information Processing Systems*, 36:71242–71262, 2023.

[25] M Whiticar, L Christensen, C Salas, and P Reece. GHGMap: novel approach for aerial measurements of greenhouse gas emissions British Columbia. *Geoscience BC Summary of Activities 2017: Energy, Geoscience BC, Report 2018-4*, pages 1–10, 2018.

[26] M Whiticar, L Christensen, C Salas, and P Reece. GHGMap: Detection of fugitive methane leaks from natural gas pipelines British Columbia and Alberta. *Geoscience BC Summary of Activities 2018: Energy and Water, Geoscience BC, Report 2019-2*, pages 67–76, 2019.

[27] M Whiticar, D Hollenbeck, B Billwiller, C Salas, and L.E Christensen. Application of the BC GHGMapper™ platform for the Alberta Methane Field Challenge (AMFC). *Geoscience BC Summary of Activities 2019: Energy and Water, Geoscience BC, Report 2020-02*, pages 87–102, 2020.

Index

For Product Safety Concerns and Information please contact our EU
representative GPSR@taylorandfrancis.com
Taylor & Francis Verlag GmbH, Kaufingerstraße 24, 80331 München, Germany

www.ingramcontent.com/pod-product-compliance
Lightning Source LLC
Chambersburg PA
CBHW082006190326
41458CB00010B/3094